Indoor Sound Environment and Acoustic Perception

Intellectual Property Publishing House

Beijing, China

Indoor Sound Environment and Acoustic Perception
ISBN 978 – 7 – 5130 – 7720 – 5
Copyright © 2020 by Qi Meng and Yue Wu
Published by Intellectual Property Publishing House
No. 50, Qixiang Road, Haidian District, Beijing, P. R China, 100081

www.cnipr.com

All rights reserved. No part of this publication may be reproduced, stored in a retrieval system, or transmitted, in any form or by any means, electronic, mechanical, photocopying, recording, or otherwise, without the prior permission of Intellectual Property Publishing House.

健康声环境营造

INDOOR SOUND ENVIRONMENT AND ACOUSTIC PERCEPTION

室内声环境与声感知

孟琪 武悦 著

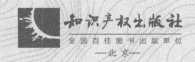
全国百佳图书出版单位
—北京—

图书在版编目（CIP）数据

室内声环境与声感知=Indoor Sound Environment and Acoustic Perception：英文/孟琪，武悦著． —北京：知识产权出版社，2020.9
ISBN 978-7-5130-7220-5

Ⅰ.①室… Ⅱ.①孟… ②武… Ⅲ.①室内声学—研究—英文 Ⅳ.①TU112

中国版本图书馆 CIP 数据核字（2020）第 188460 号

责任编辑：张　冰　　　　　　　　责任校对：谷　洋
封面设计：杰意飞扬·张　悦　　　　责任印制：孙婷婷

室内声环境与声感知
Indoor Sound Environment and Acoustic Perception
孟琪　武悦　著

出版发行：	知识产权出版社有限责任公司	网　址：	http://www.ipph.cn
社　址：	北京市海淀区气象路 50 号院	邮　编：	100081
责编电话：	010-82000860 转 8024	责编邮箱：	740666854@qq.com
发行电话：	010-82000860 转 8101/8102	发行传真：	010-82000893/82005070/82000270
印　刷：	北京建宏印刷有限公司	经　销：	各大网上书店、新华书店及相关专业书店
开　本：	787mm×1092mm　1/16	印　张：	17
版　次：	2020 年 9 月第 1 版	印　次：	2020 年 9 月第 1 次印刷
字　数：	185 千字	定　价：	168.00 元
ISBN 978-7-5130-7220-5			

出版权专有　侵权必究
如有印装质量问题，本社负责调换。

Perface

In the past two decades of the 21st century, with the improvement of environmental awareness, the demands for creating comfortable acoustic environment are increasing as well. However, with the increase of building equipment and high-density crowd activities in indoor spaces, the acoustic environment of indoor space becomes more and more complex and difficult to control, which may not only affect the efficiency of work or mental health of users, but also bring about the evacuation problems of public address (PA) system. With the development of green building and healthy building, there have been many major new developments in the field of indoor sound environment in terms of research and practice. Whilst noise mapping software have been developed and applied extensively in practice with the advancement of computing resources. Correspondingly, there have been a series of new noise control measures and design methods. In the subjective aspect, several evaluation methods have been developed with multidisciplinary approach. In the meantime, the importance of indoor sound perception design has been widely recognized, which is a major step forward from simply reducing indoor noise level. In terms of environmental policies and regulations, noise problems have been paid great attention at various levels, especially in China, leading to a series of substantial actions in noise abatement.

The main motivation of this book is to present the state-of-the-art development in indoor sound environment. This book does not simply repeat the study of indoor building acoustic environment or noise level, but considers the differences in the use of different buildings and the perception of acoustic environment, as well as the relationship between sound environment quality, health and behavior. The book involves the multidisciplinary studies in architecture, acoustics, environmental science, psychology, sociology and management. It can be used as a guide for government departments, developers, planners and architects to understand the effects of architectural design on acoustic environment.

The book is divided into five chapters, which are acoustic perception of commercial spaces, dining spaces, railway stations, hospitals, and living spaces. It involves the influences of acoustic environment on the customer satisfaction in commercial building, the diner behavior in dining spaces, the speech intelligibility in traffic buildings, the patients' psychophysiology in hospitals, and the residents' sleep in residential buildings. Chapter 1 briefly introduces the sound environment and acoustic perception in commercial spaces, including the influence of individual sound sources and user's social and behavioral characteristics on the subjective loudness and acoustic comfort in commercial spaces, and uses artificial neural network to predict. Chapter 2 discusses the sound environment and acoustic perception in dining spaces, including the effects of typical dining styles on conversation behaviors and acoustic perception, and the impact of children on the sound environment in fast-food restaurants. Chapter 3 focuses on the sound environment and acoustic perception in railway stations. This includes the acoustic environment and acoustic comfort of railway station, and the complexity of sound environment contributing to acoustic comfort in urban intermodal transit spaces. Chapter 4 describes a series of researches on sound environment and acoustic perception in hospitals, including the interaction between sound and thermal influences on patient comfort in the hospitals of China's northern heating region, and the influence of the acoustic environment in hospital wards on patient physiological and psychological indices. Chapter 5 presents the sound environment and acoustic perception in living spaces, especially the acoustic environment on the sleep of college students and elder adults.

The researches in this book is funded by the National Natural Science Foundation of China (NSFC) (Project Number: 51878210, 51808160, 51678180, 51608147, 51308145 and 50928801), the PR China Ministry of Education Foundation for PhD Bases (20112302110045), the Fundamental Research Funds for the Central Universities (HIT. NSRIF. 2020035), the Open Projects Fund of Key

Laboratory of Ecology and Energy-saving Study of Dense Habitat (Tongji University), the Ministry of Education (2020030103), the Harbin Technological Innovation Fund (2012RFXXS046), and the Natural Science Foundation of the Heilongjiang Province (YQ2019E022). Express the authors' appreciation to Prof. Jian Kang for the useful suggestion, help and support. The authors would like to thank Shilun Zhang, Shanshan Liu, Shuishan Zhao, Wenzhong Zheng, Yongxiang Wu, Lei Li, Jingyi Mu, Tianfu Zhou and Jingwen Zhang for the assistant to the survey work, and Ligang Shi for useful discussion. The authors also want to thank the Internal Medicine Department of the First Hospital of Harbin and the Second and Fourth Affiliated Hospitals of Harbin Medical University in Harbin, for their permission for sound scape investigation, respectively.

Contents

Chapter 1   Sound Environment and Acoustic Perception in Commercial Spaces

1.1   Effects of individual sound sources on the subjective loudness and acoustic comfort in underground shopping streets ·················· 3

1.2   Influence of social and behavioural characteristics of users on their evaluation of subjective loudness and acoustic comfort in shopping malls ·················· 16

1.3   Prediction of subjective loudness in underground shopping streets using artificial neural network ·················· 38

Chapter 2   Sound Environment and Acoustic Perception in Dining Spaces

2.1   Effects of typical dining styles on conversation behaviours and acoustic perception in restaurants ·················· 61

2.2   Effect of children on the sound environment in fast-food restaurants ·················· 87

Chapter 3   Sound Environment and Acoustic Perception in Railway Stations

3.1   Acoustic comfort in large railway stations ·················· 115

3.2   The complexity of sound environment contributing to acoustic comfort in urban intermodal transit spaces ·················· 144

Chapter 4   Sound Environment and Acoustic Perception in Hospitals

4.1   Interaction between sound and thermal influences on patient comfort in the hospitals of China's northern heating region ············ 161

4.2 Influence of the acoustic environment in hospital wards on patient physiological and psychological indices ································ 186

Chapter 5 Sound Environment and Acoustic Perception in Living Spaces

5.1 Effects of sound environment on the sleep of college students ······ 219

5.2 Effects of traffic noise on the sleep of elder adults ····················· 250

Chapter 1

Sound Environment and Acoustic Perception in Commercial Spaces

1.1 Effects of individual sound sources on the subjective loudness and acoustic comfort in underground shopping streets

1.1.1 Introduction

Underground shopping streets are very common in China, and acoustic problems have become one of the major research topics as a result of the construction of underground shopping streets in recent years (Wang, 2000). Although several studies on reducing noise in underground shopping streets (Tong, 1996; CAE, 2001; Ding, 2008) have been conducted, other studies suggest that the human evaluation of subjective loudness and acoustic comfort depends on a series of factors, such as individual sound sources, social characteristics of users, and general environment, rather than on sound levels only (Schafer 1977; Gaver, 1993; Dubois, 2000; Yang and Kang, 2005; Mao, 2009). In the current study, several main sound sources, such as background music, music from shops, public address (PA) system, and vendor shouts in underground shopping streets, are analysed to determine their influence on subjective loudness and acoustic comfort.

1.1.2 Methodology

A field study was conducted through a questionnaire survey at selected case study sites in Harbin, China. The locations for conducting the survey were in five typical underground shopping streets, namely, Shitoudao underground shopping street (Shitoudao), Railway station underground shopping street (Railway station), Qiulin underground shopping street (Qiulin), Huizhan underground shopping street

(Huizhan), and Lesong underground shopping street (Lesong). In terms of subjective investigation, more than 2,800 valid questionnaires were obtained from winter of 2007 to autumn of 2008 in these underground shopping streets. Around 400 to 600 interviews were conducted at each site using the same questionnaire. The interviewees in all the field surveys were randomly selected and their educational and social backgrounds as well as on-site behaviours were proven to be representative (Meng, 2010). The correlations between these factors and subjective loudness as well as acoustic comfort are shown in Table 1.1. The surveys covered four seasons, and were conducted at varying time from morning to evening, which were separated into three periods (09:00 to 11:59, 12:00 to 14:59, and 15:00 to 18:00). A five-point bipolar category scale was used in the questionnaire design. Subjective loudness was divided into five levels: 1, very quiet; 2, quiet; 3, neither quiet nor noisy; 4, noisy; and 5, very noisy. Acoustic comfort was divided into five levels: 1, very uncomfortable; 2, uncomfortable; 3, neither comfortable nor uncomfortable; 4, comfortable; and 5, very comfortable. Before the formal investigation was conducted, questionnaire reliability and validity were tested for the suitability of the final questionnaire (Meng, 2010). Before the questionnaire survey, the interviewees were told, spending one or two minutes, to evaluate the subjective loudness and acoustic comfort. Considering the interviewees need a period of 30 minutes to appropriate the sound environment in the spaces (Kang, 2004), the users who are in the underground shopping streets less than half an hour were not interviewed. The survey locations were distributed evenly in every underground shopping street to ensure that typical areas were considered, from them two underground shopping streets were discussed in this paper, including Shitoudao and Railway Station, Fig. 1.1 shows their plans. The surveys were conducted every 10 minutes in these survey locations to ensure stochastic behaviour in the survey.

Chapter 1 Sound Environment and Acoustic Perception in Commercial Spaces

Table 1.1 The correlations between other factors and subjective loudness as well as acoustic comfort

Variables		Subjective loudness					Acoustic comfort				
		Shitou dao	Railway Station	Qiulin	Huizhan	Lesong	Shitou dao	Railway station	Qiulin	Huizhan	Lesong
Interviewees' social backgrounds	Gender	0.05	-0.24*	0.13	0.01	0.00	-0.19*	-0.15	-0.05	-0.07	0.03
	Age	-0.06	0.08	0.10	-0.01	-0.02	-0.03	-0.01	-0.01	-0.05	0.09
	Income	0.38**	0.12**	0.21**	0.30**	0.26**	-0.45**	-0.44**	-0.52**	-0.51**	-0.37**
	Education	0.20**	0.06	0.07	0.04	0.05	-0.32**	-0.41**	-0.39**	-0.45**	-0.40**
	Occupation	0.22**	0.27**	0.19**	0.13	0.19**	0.19**	0.17	0.18*	0.16*	0.17**
Interviewees' behaviours	Aim of coming	0.15**	0.19**	0.13*	0.18**	0.17**	0.10	0.13**	0.10	0.09	0.16**
	Frequency of coming	-0.29**	-0.37**	-0.20**	-0.18**	-0.30**	0.20**	0.16**	0.18**	0.13**	0.26**
	Seasons	0.08	0.18	0.15	0.17	0.20**	0.23*	0.18*	0.19*	0.18	0.19**
	Visit time	0.04	0.02	0.02	0.03	0.05	0.09	0.11	0.04	0.08	0.10
	Stay time	-0.16**	-0.32**	-0.08	-0.14**	-0.33**	0.06	0.27**	0.22**	-0.01	0.12*
	Partners	0.03	-0.31**	0.08	0.10*	-0.04	-0.09	0.01	0.11	-0.01	-0.12
Environmental variables	Density of people	0.27**	0.30**	0.34**	0.36**	0.38**	-0.21**	-0.16**	-0.15**	-0.19**	-0.17**
	SPL	0.69**	0.64**	0.74**	0.64**	0.79**	-0.26**	-0.39**	-0.25**	-0.16**	-0.21**
	Air temperature	-0.02	0.01	-0.11**	-0.01	-0.06	0.13**	0.04	0.05	0.09	0.07
	Relative humidity	-0.10	-0.29**	0.30**	0.26**	-0.33**	0.16**	0.35**	-0.32**	-0.33**	0.23*
	Horizontal luminance	0.02	-0.07	-0.11**	-0.10*	0.23**	0.32**	0.18**	0.27**	0.23**	0.29**
	Reverberation	-0.03	0.25**	0.30**	0.03	-0.09	-0.27**	-0.23**	-0.34**	-0.32**	-0.10**

Note: ** indicates $p < 0.01$, and * indicates $p < 0.05$.

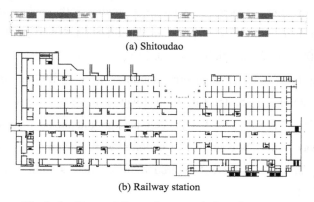

(a) Shitoudao

(b) Railway station

Fig. 1.1 Plans of the underground shopping streets

The SPSS 14.0 software was used to establish a database of all results. T-test (two-tailed) was used for factors with two scales, such as subjective loudness with any sound or without sound.

1.1.3 Results and analysis

Yu (2009) indicated that the users can define the sound environment used 5 sounds or 3 sounds they firstly heard, therefore, interviewees were asked to describe up to 5 sounds they heard in the underground shopping streets during the interview, and to note the 3 sounds they heard at first. If the sound was pointed by the interviewee in the 5 heard sounds, it was defined "with the sound", on the other hand, if the sound was not pointed by the interviewee, it was defined "without the sound".

The statistical analysis of the survey results reveals four sound sources that were cited most frequently (>300 times) by the interviewees in the questionnaire survey. These sources are background music (1,485 times), music from shops (1,077 times), PA system (942 times), and vendor shouts (880 times). This section focuses on the relationships between these sounds and subjective loudness or acoustic comfort.

1.1.3.1 Background Music

In all survey sites, subjective loudness is higher with background music than without, with a mean difference of 0.21 in Shitoudao ($p \leq 0.01$), 0.08 in Railway station ($p \leq 0.05$), 0.24 in Qiulin ($p \leq 0.01$), 0.19 in Huizhan ($p \leq 0.01$), and 0.11 in Lesong ($p \leq 0.05$). Subjective loudness is also different in every survey location. For example, in Shitoudao, the subjective loudness with background music ranges from 2.20 to 3.80, and that without background music ranges from 1.80 to 3.50. In Railway station, the subjective loudness with background music ranges from 2.00 to 3.40, and that without background music ranges from 2.10 to 3.30, as shown in Fig. 1.2.

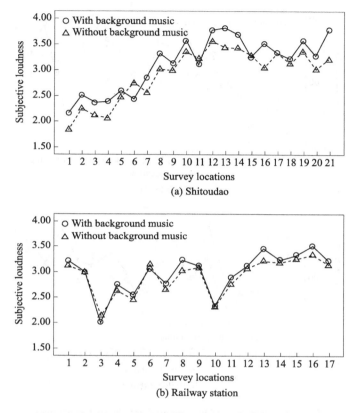

Fig. 1.2 Influence of background music on the subjective loudness evaluation in the underground shopping streets

Acoustic comfort is also higher with background music than without background music, with a mean difference of 0.32 in Shitoudao ($p \leqslant 0.01$), 0.18 in Railway station ($p \leqslant 0.05$), 0.22 in Qiulin ($p \leqslant 0.01$), 0.29 in Huizhan ($p \leqslant 0.01$), and 0.26 in Lesong ($p \leqslant 0.01$). Acoustic comfort is also different in every survey location. For example, in Shitoudao, the acoustic comfort with background music ranges from 2.30 to 3.90, whereas that without background music is from 2.20 to 3.40. In Railway station, the acoustic comfort with background music ranges from 2.40 to 3.90, whereas that without background music is from 2.60 to 3.40, as shown in Fig. 1.3.

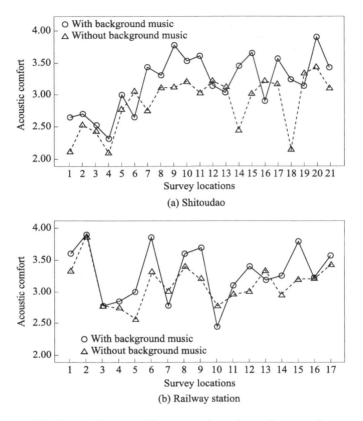

Fig. 1.3 Influence of background music on the acoustic comfort evaluation in the underground shopping streets

1.1.3.2 Music from shops

In all survey sites, subjective loudness is higher with than without music from shops, with a mean difference of 0.36 in Shitoudao ($p \leqslant 0.01$), 0.64 in Railway station ($p \leqslant 0.01$), 0.28 in Qiulin ($p \leqslant 0.01$), 0.34 in Huizhan ($p \leqslant 0.01$), and 0.40 in Lesong ($p \leqslant 0.01$). Subjective loudness also varies in every survey location. In Shitoudao, the subjective loudness with music from shops ranges from 3.20 to 3.90, whereas that without music from shops ranges from 2.70 to 3.40. In Railway station, the subjective loudness with music from shops ranges from 3.30 to 4.20, whereas that without music from shops is 2.50 to 3.40, as shown in Fig. 1.4.

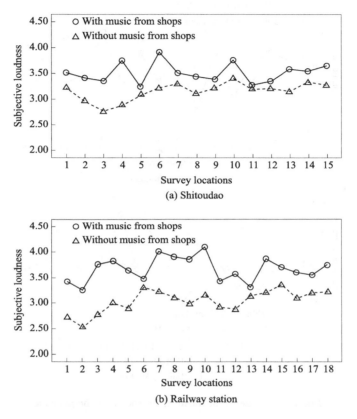

Fig. 1.4 Influence of music from shops on the subjective loudness evaluation in the underground shopping streets

Acoustic comfort, however, is lower with music from shops than that without in most survey sites, with a mean difference of −0.38 in Railway station ($p \leqslant 0.01$), −0.11 in Qiulin ($p \leqslant 0.05$), −0.20 in Huizhan ($p \leqslant 0.05$), and −0.28 in Lesong ($p \leqslant 0.05$). The mean difference of acoustic comfort is not significant in Shitoudao, and acoustic comfort lacks regularity with or without music from shops. In Railway station, the acoustic comfort with music from shops ranges from 2.40 to 3.90, whereas that without music from shops is 2.60 to 3.40, as shown in Fig. 1.5.

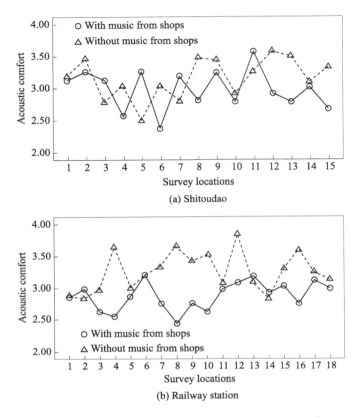

Fig. 1.5 Influence of music from shops on the acoustic comfort evaluation in the underground shopping streets

1.1.3.3 PA System

In all survey sites, subjective loudness with PA system is higher than that without, with a mean difference of 0.20 in Shitoudao ($p \leqslant 0.01$), 0.20 in Huizhan ($p \leqslant 0.01$), and 0.15 in Lesong ($p \leqslant 0.05$). It is noted that PA system did not work in Qiulin and Railway station. Subjective loudness also varies in every survey location. In Shitoudao, the subjective loudness with PA system ranges from 2.40 to 4.30, whereas that without PA system ranges from 2.60 to 3.70. In Railway station, the subjective loudness with PA system ranges from 2.60 to 4.10, whereas that without PA system ranges from 2.70 to 3.80, as shown in Fig. 1.6. The influence on

Chapter 1 Sound Environment and Acoustic Perception in Commercial Spaces

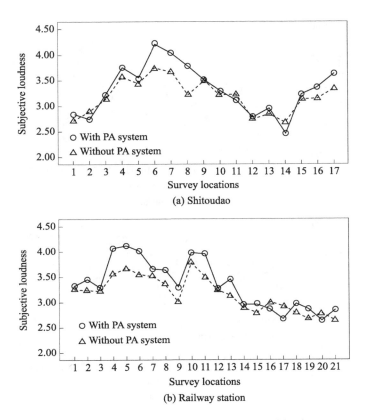

Fig. 1.6 Influence of PA system on the subjective loudness evaluation in the underground shopping streets

subjective loudness by PA system is lower than that by music from shops in several locations because the music from shops was too loud that PA system could not be heard by the interviewees.

Acoustic comfort is also higher with PA system than that without, with a mean difference of 0.27 in Shitoudao ($p \leqslant 0.01$), 0.25 in Huizhan ($p \leqslant 0.01$), and 0.21 in Lesong ($p \leqslant 0.01$). The influence on acoustic comfort by PA system is lower than that by background music, although their levels of influence on subjective loudness are nearly the same. In Shitoudao, the acoustic comfort with PA system ranges from 2.20 to 3.80, whereas that without PA system is from 2.10 to 3.60. In Huizhan, the acoustic comfort with PA system ranges from 2.60 to 3.90, whereas that without PA system is 2.90 to 3.70, as shown in Fig. 1.7. Based on the questionnaire survey,

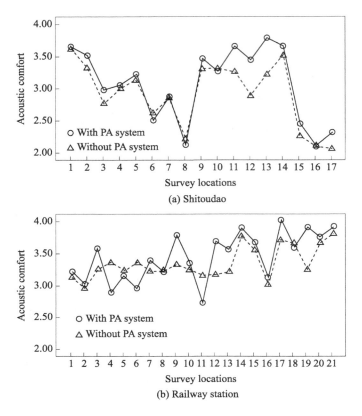

Fig. 1.7 Influence of PA system on the acoustic comfort evaluation in the underground shopping streets

many interviewees considered PA system helpful in obtaining shopping information, whereas others considered PA system annoying because of repetition. Therefore, a decrease in PA system's repetitive information may increase the acoustic comfort of the interviewees.

1.1.3.4 Vendor shouts

In several underground shopping streets where PA system does not work, such as Qiulin and Railway station, vendors have to shout to attract the attention of customers. In these survey sites, subjective loudness is higher with vendor shouts than without vendor shouts, as expected, with a mean difference of 0.43 in Qiulin ($p \leqslant 0.01$), and 0.29 in Railway station ($p \leqslant 0.01$). In Qiulin, the subjective loudness is 3.30 to 4.00 with vendor shouts, whereas that without vendor shouts is 2.60 to 3.60. In Railway station, the subjective loudness with vendor shouts ranges from 3.10 to 3.70, whereas that without vendor shouts is 2.60 to 3.40, as shown in Fig. 1.8.

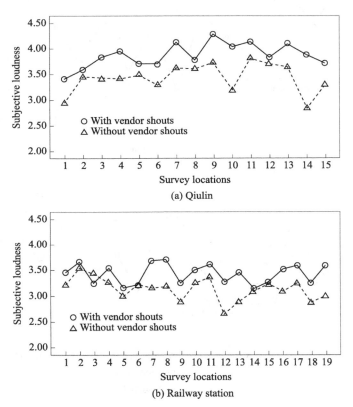

Fig. 1.8 Influence of vendor shouts on the subjective loudness evaluation in the underground shopping streets

However, the acoustic comfort vendor shouts is lower than without in most survey sites, with a mean difference of -0.62 in Qiulin ($p \leqslant 0.01$), and -0.39 in Railway station ($p \leqslant 0.01$). In Qiulin, the acoustic comfort with vendor shouts ranges from 2.20 to 3.30, whereas that without vendor shouts is 3.10 to 3.90. In Railway station, the acoustic comfort with vendor shouts ranges from 2.60 to 3.40, whereas that without vendor shouts is 3.00 to 3.70, as shown in Fig. 1.9. Comparing PA system and vendor shouts, which have the same goal of providing shopping

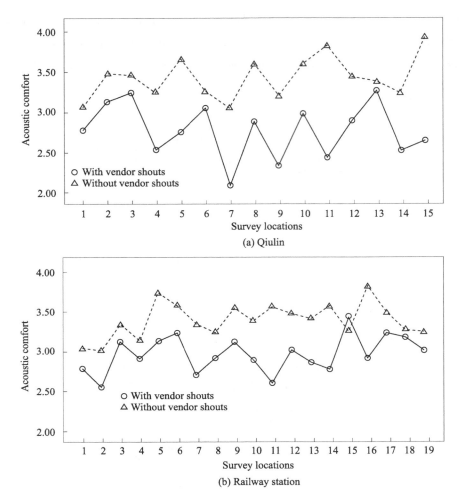

Fig. 1.9 Influence of vendor shouts on the acoustic comfort evaluation in the underground shopping streets

information, the latter is usually considered as noise, hence causing discomfort to respondents. Therefore, PA system instead of vendor shouts may be a better way to increase acoustic comfort.

1.1.4 Conclusions

Based on the analysis of the case study results, all individual sound sources can increase subjective loudness to a certain degree in underground shopping streets. However, their levels of influence on acoustic comfort are different. Background music and PA system can increase acoustic comfort, with a mean difference of 0.18 to 0.32 and 0.21 to 0.27, respectively. Music from shops and vendor shouts can decrease acoustic comfort, with a mean difference of −0.11 to −0.38 and −0.39 to −0.62, respectively.

These results can enhance knowledge on the effect of individual sound sources. With these results, acoustic comfort can be improved by changing certain sound sources.

References

[1] DING J H. A research on fire safety evacuation design in underground shopping streets [D]. Harbin: Harbin Institute of Technology, 2008.

[2] DUBOIS D. Categories as acts of meaning: the case of categories in olfaction and audition [J]. Cognitive Science Quarterly, 2002 (1): 35-68.

[3] KANG J. Urban sound environment [M]. London: Taylor and Francis, 2004.

[4] GAVER W. What in the world do we hear: an ecological approach to auditory event perception [J]. Ecological Psychology, 1993 (5): 1-29.

[5] MAO D X. Recent progress in hearing perception of loudness [J]. Technical Acoustics, 2009, 28 (6): 693-696.

[6] MENG Q. Research and prediction on soundscape in underground shopping streets [D]. Harbin: Harbin Institute of Technology, 2010.

[7] Research Teams of Chinese Academy of Engineering. Development and utilization of underground space in urban areas of China [M]. Beijing: China Architecture & Building Press, 2001.

[8] SCHAFER R M. The tuning of the world [M]. New York: Knopf, 1977.

[9] TONG L X. Planning and design in underground shopping streets [M]. Beijing: China Architecture & Building Press, 2006: 57.

[10] WANG W Q. Plan and design in underground space of urban areas [M]. Nanjing: Southeast University Press, 2000.

[11] YANG W, KANG J. Acoustic comfort evaluation in urban open public spaces [J]. Applied Acoustics, 2005, 66 (2): 211-229.

[12] YU L. Soundscape evaluation and ANN modeling in urban open spaces [D]. Sheffield: University of Sheffield, 2009.

1.2 Influence of social and behavioural characteristics of users on their evaluation of subjective loudness and acoustic comfort in shopping malls

1.2.1 Introduction

There are about 20,000 shopping malls in China[1]. Harbin City, for example, has 50 shopping malls, while 20 more will be constructed before 2030[2]. The evaluation of acoustic environment in such spaces has been paid increasing attention by users, researchers and governmental organisations [3,4]. While some studies have been carried out in terms of acoustic environment in shopping malls, the main focus has been on noise control [5] and also, only limited types of shopping mall, such as underground shopping centres, have been considered[6,7].

Previous studies suggested that the sound environment evaluation of a space depends strongly on the social characteristics of the users, such as gender, age, income, occupation, and education, as well as their behavioural characteristics [8-24]. Mehrabian [12] indicated a slight tendency for women to be more sensitive to sound than men, and evidence suggests that females generally have a higher arousal level than males[13]. In terms of age influence, Yang and Kang[14] found in their research on urban open public spaces that users are more favourable of, or tolerant towards sounds relating to nature with the increase of age. With regards to the education factor, Yu and Kang [15-17] indicated that the correlation coefficients for natural sounds are predominantly negative, suggesting that people tend to prefer natural sounds more with the increase in education level. For human sounds, mixed positive and negative correlation coefficients are found. Kang[18] indicated that the behavioural characteristics of users may also play an important role, and sound quality of an urban area may depend on how long people have been living there[19]. Similarly, Bull[20] found that people with stereos may have different sound evaluation from others. Bertoni et al. [21] also stated that sound experience is important, but Job et al. [22] obtained contradictory results. Della Crociata et al. [23-24] indicated that users' acoustic satisfaction is highly correlated with perceived acoustic intensity and is influenced by sources of acoustic annoyance. Cultural factors play an important role as well; hence, the effect of social and behavioural factors vary significantly between different countries [25,26].

Consequently, the present study is conducted to examine the influence of social and behavioural factors on evaluation of subjective loudness and acoustic comfort in different types of shopping mall, based on a series of subjective surveys. The social factors considered include gender, age, income, occupation and education; whereas the behavioural factors include the reason for visit, frequency of visit, length of stay, period of visit, and number of accompanying partners.

1.2.2 Methods

1.2.2.1 Ethics statement

This study was approved by the Degree Committee of the School of Architecture, Harbin Institute of Technology (this governing body is equipped with an ethical review board). In the questionnaire the names of individual participants were not included, and participants provided their written informed consent to participate in this study. Participants aged 17 were accompanied by their parents.

1.2.2.2 Survey sites

The field study was conducted through a questionnaire survey at selected case study sites in Harbin, China. Harbin, the capital city of Heilongjiang province, is a typical large city in China, characterised by a continental climate, with four highly distinguishable seasons. The population is about 10 million, and its economic development is at medium level in China[2].

The 30 existing shopping malls in Harbin were first divided into two groups based on their space types, that was, 12 were considered as single-space type, where the space was generally of a single volume and it was seen as one room on the plan, and 18 were multiple-space type, where the space consisted of multiple volumes and on the plan with more than one linked rooms. This division according to space types was based on previous studies stating that space types are an important factor affecting the evaluation of acoustics in indoor spaces because reverberation time (RT) and sound pressure level (SPL) may vary in different space types [18, 27]. Three shopping malls from each space type were randomly selected as final samples. The six case-study shopping malls were Qiulin, Tongji, Manhadun, Suofeiya, Jin'an, and Huizhan, among which Qiulin, Manhadun and Huizhan were single-space type, and the others were multiple-space type. Details about these shopping malls are presented in Table 1.2. It is noted that Huizhan is a typical underground shopping mall, with which comparison could be made between underground and above-ground shopping malls.

Table 1.2 Basic information of the survey sites

Sites	Size (m^2)	Sound sources	NI	Aver. SPL [dB(A)]	Aver. SL	Aver. AC
Qiulin	31,000	Background music, music from shops, PA system, footsteps, surrounding speech, air-conditioning	372	71.31	3.36	3.08
Tongji	10,000	Background music, vendors' shouting, music from shops, sounds from toys, surrounding speech, footsteps, air-conditioning	188	73.28	3.52	2.73
Manhadun	28,700	Background music, vendors' shouting, music from shops, sounds from toys, surrounding speech, footsteps, air-conditioning	302	71.43	3.48	2.96
Suofeiya	32,000	Background music, music from shops, PA system, footsteps, surrounding speech, air-conditioning	285	70.80	3.32	2.80
Jin'an	45,000	Background music, music from shops, PA system, footsteps, surrounding speech, air-conditioning	297	68.31	3.20	3.41
Huizhan	30,000	Background music, vendors' shouting, music from shops, PA system, footsteps, surrounding speech, air-conditioning, water	690	69.42	3.30	3.27

Note: The basic information includes size, sound sources, the number of interviews conducted, average SPL, average subjective loudness (1, very quiet; 2, quiet; 3, neither quiet nor loud; 4, loud; and 5, very loud) and acoustic comfort (1, very uncomfortable; 2, uncomfortable; 3, neither comfortable nor uncomfortable; 4, comfortable; and 5, very comfortable). "NI" means number of interviews, "Aver. SL" means average evaluation of subjective loudness, and "Aver. AC" means average evaluation of acoustic comfort.

1.2.2.3 Questionnaire survey

Around 100 to 700 valid questionnaires were obtained in every shopping mall from autumn 2011 to summer 2012. In total, 2,134 questionnaires were received. Previous studies suggested that 100 valid questionnaires are appropriate to evaluate the acoustics of a particular place[17,18], whereas to examine the effects of users' social or behavioural factors a considerably larger number of questionnaires would be needed [18]. The interviewees in the case study sites were randomly selected, and based on the initial data analysis, their educational and social backgrounds, as well as on-site behaviours (such as waiting for someone, walking around, shopping or passing by), were considered as representatives [22-26]. Generally 10-15 fixed survey positions were used, which were equally distributed at every survey site, and more than 20 meters from each other[18].

The questionnaire included four parts, according to the framework suggested by Kang [18], considering user, space, sound, and environment. The first part was about a user's basic information in terms of social characteristics (such as gender, age, income, education level, and occupation) and behavioural characteristics (such as what they are doing, frequency of visit, time of visit, and accompanying persons). The second part was on the evaluation of space, especially the evaluation of reverberation and space perception. The third part was about sound sources and the evaluation of loudness. The final part was the evaluation of other environmental factors, such as temperature, humidity, and lighting. The questionnaire was introduced as an enquiry on general environmental conditions, instead of concentrating solely on acoustic environment, to avoid possible bias towards the acoustic aspect[28]. The questionnaire's reliability and validity were tested before the actual field surveys[29].

A five-point bipolar category scale was used in the questionnaire. Evaluation of subjective loudness was divided into five levels: 1, very quiet; 2, quiet; 3, neither quiet nor loud; 4, loud; and 5, very loud. Evaluation of acoustic comfort was also divided into five levels: 1, very uncomfortable; 2, uncomfortable; 3, neither comfortable nor uncomfortable; 4, comfortable; and 5, very comfortable. For other questions and scales in the questionnaire, more details can be found[29].

The surveys covered four seasons because previous studies in urbanopen public spaces indicated that seasons may affect users' evaluation of acoustics [17]. Moreover, the surveys were conducted at various times of day, from morning to afternoon [14, 18]. Three time periods were considered, namely, 09:00 to 11:59 (morning), 12:00 to 14:59 (midday), and 15:00 to 17:59 (afternoon).

1.2.2.4 Objective measurements

The sound level measurement was conducted immediately after each questionnaire interview, and the microphone of the sound level meter was positioned near the location of questionnaire interview and more than 1m away from any reflective surfaces and 1.2m to 1.5m[18] above the floor to avoid the effect of sound reflections. The sound level meter was set into slow-mode, and reading was taken every 3s to 5s. A total of 100 measurement data were obtained in each survey position, and the corresponding A-weighted equivalent continuous sound level, L_{Aeq}, was derived [30-32]. In other words, after each interview a 300-500s measurement was made. Simultaneously, other environmental factors, such as air temperature, relative humidity, and lighting were also measured at the survey positions corresponding to every sound level measurement [33,34], although in this paper, due to the limitation of space, these data are not included and analysed.

1.2.2.5 Data analysis

SPSS 15.0 was used to establish a database with all the subjective and objective results [35]. The data were analysed using the following: Chi-square correlations (two-tailed) for factors with three or more categories of ranked variables; Chi-square contingency correlations (two-tailed) for factors with three or more categories for categorical variables; and mean differences t-test (two-tailed) for factors with two categories. Both linear and nonlinear correlations were considered[36].

1.2.3 Results

1.2.3.1 Social characteristics

Gender. Although previous studies suggested that the effect of gender on sound annoyance evaluations is generally insignificant[37,38], they are more focused on urban open spaces. In this study, the mean difference in evaluation of subjective loudness and acoustic comfort was determined between males and females of every survey site, as shown in Table 1.3. It is interesting to note that in shopping malls, again, no significant ($p > 0.1$) difference was observed between males and females.

Table 1.3 Relationships between social characteristics and evaluation of subjective loudness, as well as acoustic comfort

Survey sites	Gender		Age groups		Education level		Income		Occupation	
	SL	AC	SL	AC	SL	AC	SL	AC	SL	AC
Qiulin	0.08	−0.17	0.05	−0.02	0.02	−0.33**	0.22**	−0.36**	0.21**	0.17**
Tongji	0.09	−0.08	0.11	−0.01	0.04	−0.41**	0.14*	−0.42**	0.16*	0.22**
Manhadun	0.14	−0.08	0.03	0.00	0.05	−0.38**	0.15*	−0.40**	0.12*	0.20**
Suofeiya	0.06	0.00	0.06	0.07	0.03	−0.36**	0.26**	−0.37**	0.25**	0.17*
Jin'an	0.04	−0.10	0.04	−0.10	0.06	−0.46**	0.32**	−0.51**	0.18**	0.21**
Huizhan	0.01	−0.07	−0.01	−0.05	0.04	−0.45**	0.30**	−0.47**	0.13*	0.16*

Note: The table includes mean difference between males and females, chi-square test correlation coefficients for age groups, income, education level, and chi-square test contingency coefficients for occupation, where the significance levels are also shown, with ** indicating $p < 0.01$, and * indicating $p < 0.05$. SL represents evaluation of subjective loudness, and AC represents evaluation of acoustic comfort.

Age groups. Previous studies indicated that different age groups may have different evaluations of the sound environment and sound preference, possibly because of their long-term experience[19, 21, 39-40]. In this study, the users' age were divided into 7 groups, namely, ≤17, 18 to 24, 25 to 34, 35 to 44, 45 to 54, 55 to 64, and ≥65[17]. The relationships between age groups and evaluation of subjective loudness and acoustic comfort are shown in Table 1.3. It can be seen that there is no significant difference ($p > 0.1$) among the age groups in terms of evaluation of subjective loudness or acoustic comfort. Table 1.4 further examines whether there are some tendencies that certain age groups would rate the evaluation of subjective loudness and acoustic comfort more extremely. It seems that no such tendencies are observed.

Table 1.4 Differences among age groups in terms of evaluation of subjective loudness and acoustic comfort

Survey sites	Very loud	Very quiet	Very comfortable	Very uncomfortable
Qiulin	≤17	35 – 44	≥65	25 – 34
Tongji	25 – 34	45 – 54	55 – 64	≤17
Manhadun	≤17	25 – 34	18 – 24	55 – 64
Suofeiya	45 – 54	18 – 24	18 – 24	35 – 44
Jin'an	35 – 44	55 – 64	≤17	45 – 54
Huizhan	≥65	≤17	18 – 24	35 – 44

Education level. In China, education level can be divided into 6 groups, namely, (1) non-literate, (2) primary school, (3) junior middle school, (4) senior middle school, special school or technical school, (5) college graduates, and (6) graduate or higher[34]. In this study, only three people were at the level of non-literate, so only the five other levels were considered. The relationships between education level and evaluation of subjective loudness, as well as acoustic comfort, are shown in Table 1.3. No significant ($p > 0.1$) difference was found between education level and subjective loudness, but the correlation between education level and acoustic comfort was from -0.30 to -0.50 ($p < 0.01$). In other words, the higher the users' education level is, the lower the acoustic comfort is.

Income. The users' income level was divided into 6 groups, namely, ≤1,000 yuan, 1,001 to 2,000 yuan, 2,001 to 3,000 yuan, 3,001 to 4,000 yuan, 4,001 to 5,000 yuan, and ≥5,001 yuan (1 yuan ≈ 0.15 US dollar) per month[36]. The correlations between users' income and evaluation of subjective loudness, as well as acoustic comfort, as shown in Table 1.3, where the income is measured by US dollar, are from 0.10 to 0.40 ($p < 0.01$), and from -0.30 to -0.60 ($p < 0.01$), respectively. In other words, the higher the users' income is, the higher the users' evaluation of subjective loudness is, but the lower the evaluation of acoustic comfort is.

The relationships between users' income and evaluation of subjective loudness, as well as acoustic comfort, are also illustrated in Fig. 1.10. It is interesting to note that, both evaluation of subjective loudness and acoustic comfort change significantly when the income level changes from 2,000 to 3,000 yuan per month, which is approximately the average income of people in Harbin, namely 2,450 yuan per month (≈ 367 US dollars) in 2011 and 2,518 yuan per month (≈ 378 US dollars) in 2012[41,42]. In other words, it seems that in terms of evaluation of subjective loudness and acoustic comfort, there is a significant difference between people with "lower than average income (< 2,000 yuan or < 300 US dollars)" and "higher than average income (>3,000 yuan or >450 US dollars)", with a mean difference of 0.31 to 0.52 on subjective loudness ($p < 0.01$), and of 0.49 to 0.76 on acoustic comfort ($p < 0.01$).

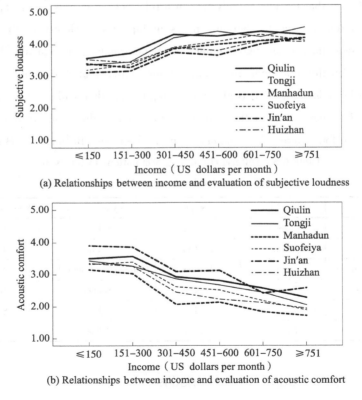

(a) Relationships between income and evaluation of subjective loudness

(b) Relationships between income and evaluation of acoustic comfort

Fig. 1.10 Relationships between users' average income and their evaluation of acoustics

Occupations. The users' occupations were divided into 12 groups, namely, farmer, soldier, worker, service worker, technical worker, teacher, officer, student, self-employed individual, housewife, retiree, and unemployed[43]. The relationships between users' occupations and evaluation of subjective loudness and acoustic comfort are shown in Table 1.3, showing Chi-square contingency correlation of 0.10 to 0.30 ($p < 0.05$) and 0.10 to 0.30 ($p < 0.05$), respectively. In other words, with different occupations, evaluation of subjective loudness and acoustic comfort also differ.

The occupations that corresponded to extreme evaluation of subjective loudness and acoustic comfort are shown in Table 1.5. It can be seen that farmers tend to select 'very loud' for evaluation of subjective loudness, and their evaluation of subjective loudness is 0.06 to 0.12 higher than that of the other occupations ($p < 0.01$) in four out of six survey sites, namely Qiulin, Tongji, Manhadun and Huizhan. Given that farmers usually live in a relatively quiet environment, they may feel nosier when they go to shopping malls in big cities. However, they may not necessarily feel the acoustic environment is unconformable, as can also be seen in Table 1.5. In the questionnaire, one farmer stated that "I would like to be surrounded by noisy crowd" It is also interesting to note that soldiers may have higher evaluation of acoustic comfort than others, with a mean difference of 0.02 to 0.07 ($p < 0.05$) in four survey sites, namely Qiulin, Suofeiya, Jin'an, and Huizhan.

Table 1.5 Differences among occupations in terms of evaluation of subjective loudness and acoustic comfort

Survey sites	Very loud	Very quiet	Very comfortable	Very uncomfortable
Qiulin	Farmer	Worker	Soldier	Teacher
Tongji	Farmer	Student	Retiree	Farmer
Manhadun	Farmer	Worker	Self-employed individual	Farmer
Suofeiya	Retiree	Officer	Soldier	Retiree
Jin'an	Technical worker	Soldier	Soldier	Student
Huizhan	Farmer	Soldier	Soldier	Retiree

1.2.3.2 Behavioural characteristics

Reason for visit. Users generally have four purposes for coming to the survey

Chapter 1 Sound Environment and Acoustic Perception in Commercial Spaces

sites, namely, shopping, walking, passing by, and waiting for someone. The relationships between the reason for visit and evaluation of subjective loudness and acoustic comfort are shown in Table 1.6. The contingency correlation between the reason for visit and evaluation of subjective loudness and acoustic comfort was from 0.10 to 0.20 ($p<0.05$) and from 0.10 to 0.30 ($p<0.05$), respectively. In other words, the users' evaluation of acoustic evaluation may be influenced by their different reasons for visit to the shopping malls.

Table 1.6 Relationship between behavioural characteristics and evaluation of subjective loudness, as well as acoustic comfort

Survey sites	Reason for visit		Frequency of visit		Time of visit						Accompanying persons	
					Seasons		Period of visit		Length of stay			
	SL	AC	SL	AC	SL	AC	SL	AC	SL	AC	SL	AC
Qiulin	0.13**	0.21**	-0.26**	0.22**	0.03	0.21**	0.04	0.02	-0.24**	0.21**	0.02	0.06
Tongji	0.17**	0.16*	-0.30**	0.19**	0.01	0.16*	0.06	0.00	-0.16**	0.26**	0.06	0.02
Manhadun	0.16*	0.14*	-0.23**	0.24**	0.07	0.12*	0.03	0.03	-0.33**	0.13*	0.01	0.07
Suofeiya	0.18**	0.19*	-0.33**	0.18**	0.05	0.20**	0.07	0.04	-0.27**	0.20**	0.04	0.01
Jin'an	0.11*	0.24**	-0.27**	0.26**	0.09	0.17*	0.02	0.04	-0.22**	0.18**	0.07	0.01
Huizhan	0.18**	0.09	-0.18**	0.13**	0.17	0.18*	0.03	0.08	-0.14**	-0.01	0.01	0.03

Note: The table includes mean difference between persons with partners and without, chi-square test correlation coefficients for frequency of visit, income, education level, and chi-square test contingency coefficients for reason for visit, where the significance levels are also shown, with ** indicating $p<0.01$, and * indicating $p<0.05$. SL represents evaluation of subjective loudness, and AC represents evaluation of acoustic comfort.

The average evaluation of subjective loudness and acoustic comfort with different reasons for visit in all survey sites are shown in Fig. 1.11. It can be seen that the users who came for shopping generally have lower evaluation of subjective loudness but higher evaluation of acoustic comfort, with a mean difference of 0.11 to 0.45 ($p<0.01$) and 0.15 to 0.58 ($p<0.01$) respectively, compared with other reasons. Conversely, the users who came to wait for someone generally have higher evaluation of subjective loudness, but lower evaluation of acoustic comfort, with a mean difference of 0.13 to 0.51 ($p<0.01$) and 0.09 to 0.49 ($p<0.05$ or $p<0.01$), respectively, compared with other purposes. This is likely because the users who are concentrating on shopping may pay decreased attention to the acoustic environment;

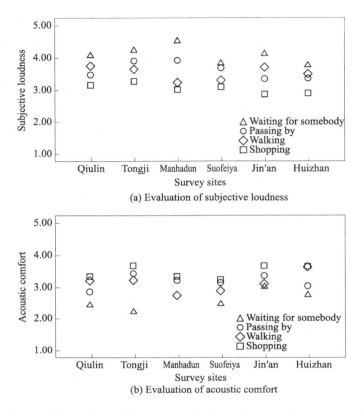

Fig. 1. 11   Effects of users' reason for visit on their evaluation of acoustics

whereas those who are waiting for someone could have increased attention.

Frequency of visit. The users' frequency of visit to a certain place and its influence on their evaluation of subjective loudness or acoustic comfort has been mentioned in previous studies[25, 37]. In this study, the users' frequency of visit to shopping malls was divided into 5 categories, namely, first time, rarely (at least once a year), sometimes (at least once a month), often (at least once a week), and frequently (more than three times a week). The relationships between the frequency of visit and evaluation of subjective loudness and acoustic comfort are shown in Table 1. 6. It can be seen that both evaluation of subjective loudness and acoustic comfort are influenced by the frequency of visit, with correlation of -0.10 to -0.40 ($p < 0.01$) in evaluation of subjective loudness, and 0. 10 to 0. 30 in evaluation of acoustic comfort, respectively. In other words, with the increase in the frequency of visit, the users' evaluation of subjective loudness is lower, but the evaluation of acoustic comfort is higher. This suggests that when users are more familiar with the environment of a certain space, such as shopping malls, they would have a better acoustic evaluation.

Fig. 1.12 presents the average evaluation of subjective loudness and acoustic comfort in all survey sites with increasing frequency of visit. It is interesting to note that in terms of evaluation of subjective loudness, there is a significant change from "once a year" to "once a month" ($p < 0.01$), by 0.52 to 0.81 in different sites. Meanwhile, the change in acoustic comfort is also significant, by 0.58 to 1.17 ($p < 0.01$).

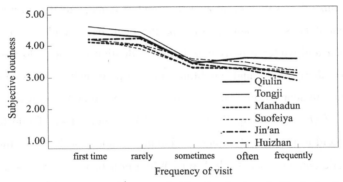

(a) Relationships between frequency of visit and evaluation of subjective loudness

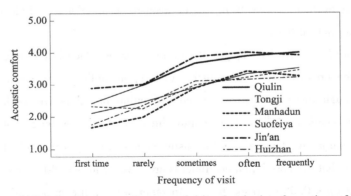

(b) Relationships between frequency of visit and evaluation of acoustic comfort

Fig. 1.12 Relationships between users' frequency of visit and their evaluation of acoustics

Note: rarely—at least once a year; sometimes—at least once a month; often—at least once a week; frequently—more than three times a week.

Time of visit. In this study, "time" was considered at three levels, namely seasons (from autumn to summer), period of visit (morning, midday, and afternoon); and length of stay (less than 1 hour, 1 to 2 hours, 2 to 3 hours, 3 to 4 hours, and more than 4 hours).

(a) *Seasons.* The relationships between seasons and evaluation of subjective loudness and acoustic comfort are shown in Table 1.6. No significant difference ($p > 0.1$) was observed between seasons in terms of evaluation of subjective loudness, but significant differences were noted between seasons in terms of evaluation of acoustic comfort, with contingency correlation coefficients of 0.10 to 0.20 ($p < 0.05$). This is perhaps because the evaluation of acoustic comfort is more related to the evaluation of other aspects of physical comfort, such as the heat and humidity conditions, and users in summer and winter usually give better evaluation for the indoor environment due to more comfortable thermal conditions in shopping malls[44], as one customer stated: "I prefer to shop in the malls in summer because they have air conditioning."

(b) *Period of visit.* No significant ($p > 0.1$) correlation was found between the period of visit and evaluation of subjective loudness, and also between the period of visit and acoustic comfort (see Table 1.6). While in urban open spaces the period of visit was found to be influential on the soundscape evaluation[45,46], perhaps because better lighting evaluation may bring better acoustic evaluation, in indoor shopping malls this may not be the case.

(c) *Length of stay.* The length of stay in a certain place is considered to have some influence on evaluation of acoustic comfort, as indicated in several previous studies, although those studies concerned evaluation of subjective loudness rather than acoustic comfort [47,48]. The relationships between the length of stay and evaluation of subjective loudness and acoustic comfort are given in Table 1.6, where it can be seen that the correlation coefficients are -0.10 to -0.40 ($p < 0.01$), and 0.10 to 0.30 ($p < 0.05$), respectively. In other words, the users will have lower evaluation of subjective loudness and higher acoustic comfort when their length of stay is longer.

The relationships between the length of stay and evaluation of subjective loudness and acoustic comfort are also shown in Fig. 1.13. It can be seen that in terms of

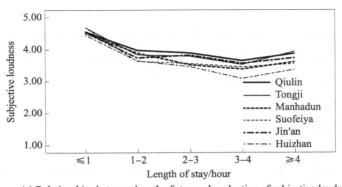

(a) Relationships between length of stay and evaluation of subjective loudness

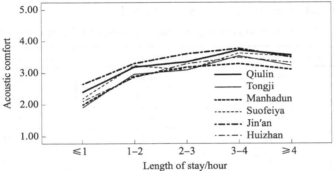

(b) Relationships between length of stay and evaluation of acoustic comfort

Fig. 1.13 Relationships between users' length of stay and their evaluation of acoustics

evaluation of subjective loudness there is a significant decrease from less than 1 hour to 1-2 hours, with a mean difference of 0.52 to 0.98 ($p < 0.01$) at different sites, whereas in terms of evaluation of acoustic comfort there is a significant increase, with a mean difference of 0.37 to 0.86 ($p < 0.01$). Such changes in evaluation of subjective loudness and acoustic comfort continue with increasing length of stay until 3-4 hours, although the rate of change is much slower. Conversely, when the length of stay is greater than 4 hours, evaluation of subjective loudness will increase and the acoustic comfort will decrease. Similar results were also obtained in a previous study[27].

Accompanying persons. Previous studies in urban open public spaces suggested that the acoustic evaluation might be different with accompanying persons[17-18, 29, 49]. In this study, however, no significant difference was obtained ($p > 0.1$) between the users with accompanying persons and without, as shown in Table 1.6, in terms of evaluation of subjective loudness as well as acoustic comfort. A possible reason for this difference is that the users staying outdoors may pay more attention to their partners, compared to indoor spaces like shopping malls.

1.2.4 Discussions

Previous studies noted that there are interrelationships between social characteristics [50,51]. This is also the case in this study, in terms of income, education level and occupation, as shown in Table 1.7.

Table 1.7 Relationships among income, education level and occupation

Survey sites	Income and education level	Occupation and income	Occupation and education level
Qiulin	0.43**	0.30**	0.26**
Tongji	0.50**	0.35**	0.32**
Manhadun	0.38**	0.28**	0.33**
Suofeiya	0.32**	0.26**	0.24**
Jin'an	0.42**	0.32**	0.36**
Huizhan	0.60**	0.37**	0.30**

Note: The table shows chi-square test correlation coefficients between income and education level, and chi-square test contingency coefficients between occupation and income as well as education level, where the significance levels (2-tailed) are also shown, with ** indicating $p<0.01$, and * indicating $p<0.05$.

Relationships between occupation and evaluation of acoustic comfort are generally insignificant when income or education level is in a certain range. For example, considering the income "from 1,000 yuan (150 US dollars) to 2,000 yuan (300 US dollars)" or the education level of "graduate or higher", the results in Table 1.8 indicate that generally there is no significant correlation between occupation and evaluation of acoustic comfort. Meanwhile, for a given occupation, such as "worker", there are significant correlations between income and evaluation of acoustic comfort, with correlation coefficients of −0.30 to −0.60, as well as between education level and evaluation of acoustic comfort, with correlations of −0.30 to −0.50, as shown in Table 1.9.

Table 1.8 Relationships between occupation and evaluation of acoustic comfort

Survey sites	Occupation and evaluation of acoustic comfort with income fixed	Occupation and evaluation of acoustic comfort with education level fixed
Qiulin	0.08	0.11
Tongji	0.10	0.05
Manhadun	0.05	0.10
Suofeiya	0.12	0.08
Jin'an	0.04	0.13 *
Huizhan	0.16	0.11

Note: The table shows chi-square test contingency coefficients between occupation and evaluation of acoustic comfort, when income or education is fixed at a level, namely income is from 151 to 300 US dollars, education level is graduate or higher, where the significance levels (2-tailed) are also shown, with ** indicating $p < 0.01$, and * indicating $p < 0.05$.

Table 1.9 Relationships between users' evaluation of acoustic comfort and income, as well as education level

Survey sites	Income	Education level
Qiulin	-0.40 **	-0.36 **
Tongji	-0.45 **	-0.42 **
Manhadun	-0.41 **	-0.40 **
Suofeiya	-0.39 **	-0.38 **
Jin'an	-0.53 **	-0.50 **
Huizhan	-0.49 **	-0.47 **

Note: The table shows chi-square test correlation coefficients between income or education level and evaluation of acoustic comfort, when occupation is fixed as worker, where the significance levels (2-tailed) are also shown, with ** indicating $p < 0.01$, and * indicating $p < 0.05$.

A multiple regression analysis[52] was then carried out to identify whether income or education level is more important for the evaluation of acoustic comfort. Table 1.10 shows that the R_{adj2}, ranged from 0.263 to 0.377, is significant ($p < 0.001$) in all survey sites. Income is found to be generally more important than education level with regards to influencing the evaluation of acoustic comfort because it has higher absolute values, from 0.047 to 0.233, of standardised coefficient in all survey sites.

Table 1.10 The effect of users' income and education on evaluation of acoustic comfort

Survey sites	R_{adj2}	Factor	Standardised coefficient
Qiulin	0.312	Income	-0.332 **
		Education level	-0.285 **
Tongji	0.287	Income	-0.410 **
		Education level	-0.207 **
Manhadun	0.365	Income	-0.338 **
		Education level	-0.109 **
Suofeiya	0.263	Income	-0.378 **
		Education level	-0.145 **
Jin'an	0.377	Income	-0.350 **
		Education level	-0.202 **
Huizhan	0.253	Income	-0.331 **
	0.253	Education level	-0.254 **

Note: The table shows multiple regression analysis R_{adj2} with standardised coefficient between income or education level and evaluation of acoustic comfort, where the significance levels (2-tailed) are also shown, with ** indicating $p < 0.001$.

It is noted that based on the data in Table 1.3 and Table 1.6, it can be shown that the space type of shopping malls, namely single-space or multiple-space type, have no significant effect on users' evaluation of acoustics.

1.2.5 Conclusions

Based on questionnaire surveys and measurements conducted in six shopping malls in Harbin City, China, this study examines the sound environment in terms of the users' social and behavioural characteristics.

In terms of social characteristics, evaluation of subjective loudness is influenced by income and occupation, with correlation coefficients or contingency coefficients of 0.10 to 0.40. The evaluation of acoustic comfort is influenced by income and education level, with correlation coefficients or contingency coefficients of 0.10 to 0.60. The effect of gender and age on the evaluation of subjective loudness and acoustic comfort is statistically insignificant.

Chapter 1 Sound Environment and Acoustic Perception in Commercial Spaces

In terms of behavioural characteristics, evaluation of subjective loudness is influenced by the reason for visit, frequency of visit, and length of stay, with correlation coefficients or contingency coefficients of 0.10 to 0.40. Evaluation of acoustic comfort is influenced by the reason for visit, frequency of visit, length of stay, and season, with correlation coefficients of 0.10 to 0.30. The users who were waiting for someone were found to give lower evaluation of acoustic comfort compared to those who were shopping; the users who went to shopping malls more than once a month were found to have higher evaluation of acoustic comfort; and the users who stayed in shopping malls from 2 to 4 hours were likely to give better evaluation of acoustic comfort compared to those who stayed longer or shorter.

Between different space types of shopping malls it seems that there is no significant difference in terms of acoustic evaluation.

The findings of this study can contribute to a better understanding of acoustic environment in shopping malls, and are also useful for the establishment of acoustic comfort prediction models based on artificial neural networks (ANN) and support vector machine (SVM), which are currently being developed. It is also of great interest to compare different kinds of shopping malls, using the same methodology. Correspondingly, a cross-cultural comparison between the UK and China is planned[53]. Finally, it is important to relate the evaluation of acoustic comfort to the sound fields of shopping malls, where the space forms are often special such as long or flat spaces, for which much theoretical work has been carried out[54-60].

References

[1] SONG F. Sound environment in comprehensive shopping malls [D]. Harbin: Harbin Institute of Technology, 2011.

[2] ZHAO M D. The Eastern Moscow: Harbin City [M]. Harbin: Harbin Press, 2009.

[3] Ministry of Environmental Protection of the People's Republic of China. Emission Standard for Community Noise: GB 22337—2008 [S]. Beijing: Chinese Environmental Science Press, 2008.

[4] General Administration of Quality Supervision, Inspection and Quarantine of the People's Republic of China. Environmental Quality Standard for Noise: GB 3096—2008 [S]. Beijing: Chinese Environmental Science Press, 2008.

[5] DENG S C. How to control the noise in shopping malls [J]. Chinese Public Health, 1994 (2): 93.

[6] TANG Z Z, KANG J, JIN H. Sound field and soundscape in underground shopping streets [R]. Proceedings of Euronoise 2009. Edinburgh, Scotland, UK, 2009.

[7] TANG Z Z, JIN H, KANG J. Acoustic environment in underground shopping streets in north China [J]. Journal of Harbin Institute of Technology, 2011, 18 (sup2): 354-359.

[8] ZIMER K, ELLERMEIER W. Psychometric properties of four measures of noise sensitivity: a comparison [J]. Journal of Environmental Psychology, 1999, 19: 295-302.

[9] KARLSSON H. The acoustic environment as a public domain [J]. The Journal of Acoustic Ecology, 2000, (1): 10-13.

[10] ELLERMEIER W, EIGENSTETTER M, ZIMMER K. Psychoacoustic correlates of individual noise sensitivity [J]. The Journal of the Acoustical Society of America, 2001, 109 (4): 1464-1473.

[11] WOOLLEY H. Urban open spaces [M]. London: Spon Press, 2003.

[12] MEHRABIAN A. Public places and private spaces—the psychology of work, play, and living environments [M]. New York: Basic Books Inc. Publisher, 1976.

[13] CROOME D J. Noise, building and people [M]. Oxford: Pergamon Press, 1977.

[14] YANG W, KANG J. Acoustic comfort evaluation in urban open public spaces [J]. Applied Acoustics, 2005, 66 (2): 211-229.

[15] YU L, KANG J. Modeling subjective evaluation of soundscape quality in urban open space: An artificial neural network approach [J]. The Journal of the Acoustical Society of America, 2009, 126 (3): 1163-1174.

[16] YU L, KANG J. Effects of social, demographical and behavioural factors on the sound level evaluation in urban open spaces [J]. The Journal of the Acoustical Society of America, 2008, 123 (2): 772-783.

[17] YU L, KANG J. Factors influencing the sound preference in urban open spaces [J]. Applied Acoustics 2010, 71 (7): 622-633.

[18] KANG J. Urban sound environment [M]. London: Taylor & Francis incorporating Spon, 2006

[19] SCHULTE-FORTKAMP B, NITSCH W. On soundscapes and their meaning regarding noise annoyance measurements. Proceedings of Inter-Noise, Fort Lauderdale, FL, USA, 1999.

[20] BULL M. Sounding out the city: Personal stereos and the management of everyday life [M]. London: Berg Publishers, 2000.

[21] BERTONI D, FRANCHINI A, MAGNONI M, et al. Reaction of people to urban traffic noise in Modena, Italy. Proceedings of the 6th Congress on Noise as a public Health Problem,

Noise, and Man, Nice, France, 1993.

[22] JOB RFS, HAFIELD J, CARTER N L, et al. Reaction to noise: the roles of soundscape, enviroscape and psychscape. Proceedings of Inter-Noise, Fort Lauderdale, USA, 1999.

[23] DELLA CROCIATA S, MARTELLOTTA F, SIMONE A. A measurement procedure to assess indoor environment quality for hypermarket workers [J]. Building and Environment, 2012, 47: 288-299.

[24] DELLA CROCIATA S, SIMONE A, MARTELLOTTA F. Acoustic comfort evaluation for hypermarket workers [J]. Building and Environment, 2013, 59: 369-378.

[25] ZHANG M, KANG J. Towards the evaluation, description and creation of soundscape in urban open spaces [J]. Environment and Planning B Planning and Design, 2007, 34 (1): 68-86.

[26] ZHANG M, KANG J. Subjective evaluation of urban environment: a case study in Beijing [J]. International Journal of Environment and Pollution, 2009, 39 (1): 187-199.

[27] CHEN B, KANG J. Acoustic comfort in shopping mall atrium spaces: a case study in Sheffield Meadow hall [J]. Architectural Science Review, 2004, 47 (2): 107-114.

[28] KANG J, ZHANG M. Semantic differential analysis of the soundscape in urban open public spaces [J]. Building and Environment, 2010, 45 (1): 150-157.

[29] MENG Q. Research and prediction on soundscape in underground shopping streets [D]. Harbin: Harbin Institute of Technology, 2010.

[30] KANG J, MENG Q, JIN H. Effects of individual sound sources on the subjective loudness and acoustic comfort in underground shopping streets [J]. Science of The Total Environment, 2012, 435-436: 80-89.

[31] BROWN A L. Soundscapes and environmental noise management [J]. Noise Control Engineering Journal, 2010, 58 (5): 493-500.

[32] MENG Q, JIN H, KANG J. Soundscape evaluation in different types of underground shopping streets [J]. Journal of Harbin Institute of Technology, 2011, 18 (sup2): 231-234.

[33] LIU S Z, YOSHINO H, MOCHIDA A. A measurement study on the indoor climate of a college classroom [J]. International Journal of Ventilation, 2011, 10 (3): 251-262.

[34] HUANG L, ZHU Y X, OUYANG Q, et al. A study on the effects of thermal, luminous, and acoustic environments on indoor environmental comfort in offices [J]. Building and Environment, 2012, 49 (1): 304-309.

[35] PALLANT J. SPSS survival manual [M]. 2nd ed. UK: Open University Press, 2005.

[36] LI L S. Design investigation [M]. Beijing: China Architecture & Building Press, 2006.

[37] FIELDS J M. Effect of personal and situational variables on noise annoyance in residential areas [J]. The Journal of the Acoustical Society of America, 1993, 93 (5): 2753-2763.

[38] MIEDEMA H M E, VOS H. Exposure-response relationships for transportation noise [J]. The Journal of the Acoustical Society of America, 1998, 104 (6): 3432-3445.

[39] YANG W. Soundscape in urban open public spaces: Comparison between Beijing and Sheffield [D]. Sheffield: University of Sheffield, 2000.

[40] JIN H, MENG Q, KANG J. Subjective evaluation of acoustic comfort in underground shopping streets [J]. INTER-NOISE and NOISE-CON Congress and Conference Proceedings, InterNoise09, Ottawa CANADA, 2009: 2782-2789.

[41] Harbin Statistical Bureau. Statistical yearbook of Harbin-2011 [M]. Beijing: China Statistics Press, 2011.

[42] Harbin Statistical Bureau. Statistical yearbook of Harbin-2012 [M]. Beijing: China Statistics Press, 2012.

[43] LIN N. Social capital: a theory of structure and action [M]. London: Cambridge University Press, 2001.

[44] MENG Q, JIN H, KANG J. The influence of users' behavioural characteristics on the evaluation of subjective loudness and acoustic comfort in underground shopping streets [J]. Hua Zhong Architecture, 2010, 28 (5): 90-92.

[45] NILSSON M E, BERGLUND B. Soundscape quality in suburban green areas and city parks [J]. Acta Acustica United with Acustica, 2006, 92 (6): 903-911.

[46] SZEREMETA B, ZANNIN P H T. Analysis and evaluation of soundscapes in public parks through interviews and measurement of noise [J]. Science of The Total Environment, 2009, 407 (24): 6143-6149.

[47] SOUTHWORTH M. The sonic environment of cities [J]. Environment and Behaviour, 1969, 1: 49-70.

[48] MAO D X. Recent progress in hearing perception of loudness [J]. Technical Acoustics, 2009, 28 (6): 693-696.

[49] YANG W, KANG J. Soundscape and sound preferences in urban squares [J]. Journal of Urban Design, 2005, 10 (1): 69-88.

[50] COSMAS S C. Life style and consumption patterns [J]. Journal of Consumer Research, 1982, 8: 453-455.

[51] REINGOLD D A. Inner-city firms and the employment problem of the urban poor: Are poor people really excluded from jobs located in their own neighbourhoods? [J]. Economic Development Quarterly, 1999, 13 (4): 23-37.

[52] DANIEL W W, TERRELL J C. Business Statistics, Basic Concepts and Methodology [M]. Boston: Houghton Mifflin Company, 1979.

[53] KANG J. Comparison of speech intelligibility between English and Chinese [J]. The Journal of the Acoustical Society of America. 1998, 103 (2): 1213-1216.

[54] KANG J. Reverberation in rectangular long enclosures with geometrically reflecting boundaries [J]. Acta Acustica united with Acustica, 1996, 82 (3): 509-516.

[55] KANG J. Acoustics in long enclosures with multiple sources [J]. The Journal of the Acoustical Society of America, 1996, 99 (2): 985-989.

[56] KANG J. Sound attenuation in long enclosures [J]. Building and Environment, 1996, 31 (3): 245-253.

[57] KANG J. Improvement of the STI of multiple loudspeakers in long enclosures by architectural treatments [J]. Applied Acoustics, 1996, 47 (2): 129-148.

[58] KANG J. A method for predicting acoustic indices in long enclosures [J]. Applied Acoustics, 1997, 51 (2): 169-180.

[59] KANG J. Scale modelling for improving the speech intelligibility from multiple loudspeakers in long enclosures by architectural acoustic treatments [J]. Acta Acustica united with Acustica, 1998, 84 (4): 689-700.

[60] KANG J. Reverberation in rectangular long enclosures with diffusely reflecting boundaries [J]. Acta Acustica united with Acustica, 2002, 88 (1): 77-87.

1.3 Prediction of subjective loudness in underground shopping streets using artificial neural network

1.3.1 Introduction

To resolve the conflict between fast economic growth and land shortage, large and medium-sized cities in China are trying to build and optimize the use of underground shopping streets. For example, as a type city in China, Harbin has built more than 15km^2 underground shopping streets between 2011 and 2020[1]. Therefore, in China acoustic problems in underground shopping streets have become an important research topics[2-4].

Soundscape approach has been widely accepted and explored in recent years[5-7]. Studies on soundscape have covered residential areas, urban squares, parks, and indoor environments[8-11]. These previous investigations suggest that subjective loudness is one of the important factors affecting the creation of good soundscape. Consequently, it is important in underground shopping streets if subjective loudness can be predicted at the design stage. However, to the best of the authors' knowledge, no work has been carried out in this respect. The aim of this paper is therefore to develop prediction models for subjective loudness in underground shopping streets.

Mathematical and statistical methods have been used in the past to develop models for prediction[12-18], including Bayesian network, logistic regression, and artificial neural network (ANN). ANN has been used in solving many problems in engineering processes and recently Yu and Kang[19-21] used ANN in their study on urban soundscape to predict acoustic comfort and subjective loudness, and good results have been obtained.

In this paper, therefore, ANN is used to predict subjective loudness in underground shopping streets. The paper starts with the examination of the ANN method and the establishment of a database; it then presents the results of different types of ANN models.

1.3.2 Methodology

1.3.2.1 ANN

ANN is an artificial intelligence (AI) system capable of obtaining an output function from a set of input variables of any type. It needs to be trained based on relevant database[22]. ANN introduces the use of silicon logic gates of microprocessors in computers to imitate operation of human brain neurons. The fundamental building blocks of ANN are nodes comparable to neurons, and weighted connections that can be likened to synapses in biological systems. A number of hidden layers are used to represent the relations between nodes. Nodes are simple information processing elements. The number of nodes in each hidden layer and the number of hidden layers can be defined according to the number of data points, and the complexity of the relationship existing between the input and output data. Generally, the use of more neurons can memorize and reduce the reasoning capability of ANN, so that an ANN model should contain a minimum number of neurons capable of simulating training data. The learning purpose of ANN is to update weighting. The input-output relationship can be simulated by adjusting the weighting. In other words, the network has to be trained to reproduce the input-output relationship with the optimal weighting.

The basic training process[23] includes (a) calculating outputs from input values, (b) checking outputs by comparing calculated and known values in database, and (c) adjusting the weight for each node to reduce the difference between the calculated and known values. The accuracy of the predictions of a network is quantified using the root of the mean squared error difference (RMSE) between the predicted and the known values.

The design of an ANN model requires that a suitable architecture be determined.

For a less complex network with a relatively better generalization, a back propagation neural network modeling algorithm requires the setting up of different learning parameters, such as iterations, training functions, the number of hidden layers, and the optimal number of nodes in a hidden layer.

In this study, MATLAB 7.0 was used to implement the ANN models. A back-propagated (BP) multilayer forward neural modeling system was adopted, including input layer, hidden layer, and output layer, as shown in Fig. 1.14. The back-propagation is the generalization of the Widrow-Hoff learning rule[22] to multiple-layer networks and nonlinear differentiable transfer functions, where a hyper tangent function is used in hidden layers and a sigmoid function is used in output layer. Therefore, the nonlinear relationships between input and output can be constructed in the models.

The network training is a process of continual readjustment between the weights and the threshold, in order to reduce the network error to a pre-set minimum or stop at a preset training step, and then by inputting the forecasting samples to the trained network, to obtain the prediction results.

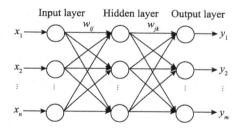

Fig. 1.14 The typical BP neural network

1.3.2.2 Database

The success of ANN models in predicting depends on the comprehensiveness of training data. A database was created through surveys and measurements in a number of case study sites in Harbin, China. Harbin is the capital city of Heilongjiang province. In the 1960s, many air-raid shelters were constructed in Harbin. In the following years some of them were transformed into underground shopping streets.

Chapter 1 Sound Environment and Acoustic Perception in Commercial Spaces

In this study, the surveys were conducted in five such underground shopping streets, namely, Shitoudao, Railway station, Qiulin, Huizhan, and Lesong. Fig. 1.15 shows their floor plans. In Table 1.11 some basic information are given on the case study sites, where a wide variation of physical conditions and space morphology was considered. Over 2,800 valid questionnaires were obtained in four seasons from the winter of 2,007 to the autumn of 2,008 from the five underground shopping streets, approximately 400-600 in each site. Interviewees in all the field surveys were randomly selected. The time of the surveys varied from morning to evening and could be roughly separated into three periods: morning (9:00-11:59), midday (12:00-14:59), and afternoon (15:00-18:00).

Table 1.11 Basic information on the five survey sites

Survey sites	Main shop types	Main sounds	Number of interviews	Floor area (m^2)	Clear height (m)	Materials		
						Floor	Wall	Ceiling
Shitoudao	Clothes	Background music, vendors' shouting, music from shops, PA system, footsteps, surrounding speech, air-conditioning	598	17,000	3.4	Marble	Gypsum board	Gypsum board
Railway station	Clothes, groceries, toys, food	Background music, vendors, shouting, music from shops, sounds from toys, surrounding speech, footsteps, air-conditioning	446	50,000	4.0	Marble	Concrete	Concrete
Qiulin	Clothes	Background music, vendors' shouting, music from shops, PA system, footsteps, surrounding speech, air-conditioning, children shouting	459	14,000	3.2	Marble	Fabric	Gypsum board
Huizhan	Clothes, groceries, food	Background music, vendors' shouting, music from shops, PA system, footsteps, surrounding speech, air-conditioning, water	690	30,000	4.2	Marble	Concrete	Gypsum board
Lesong	Clothes, groceries	Background music, vendors' shouting, music from shops, PA system, footsteps, surrounding speech, air-conditioning	629	15,000	3.4	Wood	Fabric	Glass

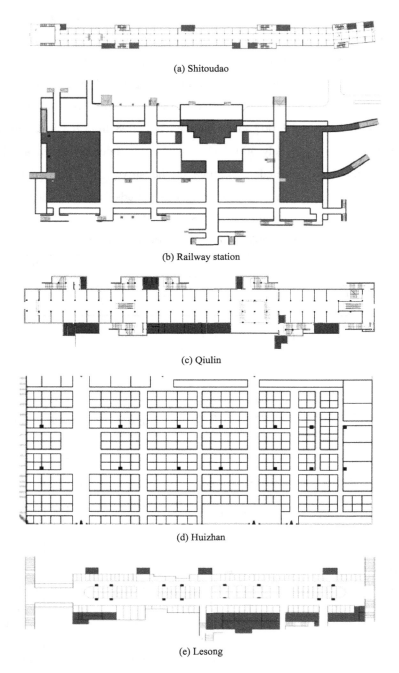

Fig. 1.15　Floor plans of survey sites

Acoustic evaluation considered various aspects affecting subjective loudness. The social-demographic factors of users were also obtained, including gender, age, occupation, education level and so on, as well as on-site behavior such as coming alone or with other people. Questions were also asked relating to the assessment of other environmental factors such as indoor temperature and humidity, taking into account possible interactions between different aspects of physical environment. For instance, a respondent under poor lighting conditions may evaluate subjective loudness differently from someone under a comfortable lighting condition. Thermal and aural interactions have also been demonstrated in numerous studies[8, 11, 24-26].

A five-point bipolar category scale was used in designing the questionnaire. For example, subjective loudness was divided into five levels: 1, very quiet; 2, quiet; 3, neither quiet nor loud; 4, loud; and 5, very loud. Each questionnaire survey was conducted within about 10 minutes to ensure that the survey sampling was distributed stochastically[27]. Before the formal investigations, reliability and validity of the questionnaire were carried out based on statistical analysis to ensure that the final questionnaire was approprite[28].

Measurement was conducted simultaneously with the questionnaire survey. Air temperature, relative humidity, horizontal luminance, sound pressure level (SPL) and density of people (persons/m^2) were measured. A one-minute equivalent continuous noise level (Leq) was recorded for each interview around the time the interviewee was filling in the questionnaire. The measurement results of the five survey sites are shown in Table 1.12.

Table 1.12 Measurement values of five survey sites

Survey sites	Air temperature (℃)			Relative humidity (%)			Horizontal luminance (lx)			SPL [dB(A)]		
	Range	Mean values	Standard deviations	Range	Mean values	Standard deviations	Range	Mean values	Standard deviations	Range	Mean values	Standard deviations
Shitoudao	11.3-25.3	18.5	3.76	58-81	72.8	5.61	60-255	105.3	32.35	53-85	71.7	5.61
Railway station	10.8-24.2	16.6	3.83	50-75	58.4	7.19	50-261	119.4	47.86	52-86	69.4	4.75

Continued

Survey sites	Air temperature (℃)			Relative humidity (%)			Horizontal luminance (lx)			SPL [dB(A)]		
	Range	Mean values	Standard deviations	Range	Mean values	Standard deviations	Range	Mean values	Standard deviations	Range	Mean values	Standard deviations
Qiulin	12.8-26.1	18.4	3.73	56-71	65.2	3.57	45-212	96.3	43.00	52-85	73.5	4.98
Huizhan	10.3-26.7	19.6	3.75	51-71	61.4	5.99	45-261	125.6	52.89	50-87	74.7	4.88
Lesong	13.3-27.2	19.5	3.74	58-80	71.1	4.02	49-230	99.4	44.66	50-86	73.0	4.96

Software SPSS 14.0 was used to establish a database with all subjective and objective results. Data analysis was performed using Pearson/Spearman correlations (two-tailed) for factors with three or more categories, and mean differences t-test (two-tailed) for factors with two categories. Both linear and nonlinear correlations were considered. Based on the significance level, 21 variables were selected as the input variables in the ANN model, with 16 objective variables and five subjective variables. In Table 1.13 the influence of these variables to subjective loudness are given, where * and * * represent $p \leqslant 0.05$ and $p \leqslant 0.01$, respectively, where p is the probability of being uncertain. The output variable is subjective loudness.

From Table 1.13 it can be seen that the positive correlation found between SPL (A1) and subjective loudness ($p < 0.01$) is generally the greatest in all objective variables. The correlations between subjective loudness and some objective variables, such as income (A4), occupation (A6), frequency of coming (A8), density of people (A13), relative humidity (A15), and horizontal luminance (A16), are also important, with significant values in at least four case sites. By contrast, gender (A2), age (A3), education (A5), aim of coming (A7), season (A9) and air temperature (A14) are less important factors in the perception of subjective loudness.

Subjective variables, including acoustic comfort evaluation (B2), humidity evaluation (B3), brightness evaluation (B4), and view assessment (B5), are usually important for subjective loudness. By contrast, evaluation of reverberation time is important for subjective loudness in only two of the survey sites. The relationship between subjective loudness and acoustic comfort evaluation is of a parabolic shape according to previous studies[24-29] and thus, the linear correlation between subjective loudness and acoustic comfort evaluation is slightly lower.

Table 1.13 Input variables in models for subjective loudness

	Variables	Measures of the attributes	Correlation or mean difference				
			Shitoudao	Railway station	Qiulin	Huizhan	Lesong
Objective variables	A1 SPL	dB(A)	0.69 **	0.64 **	0.74 **	0.64 **	0.79 **
	A2 Gender	1, male; 2, female	0.05	-0.24 *	0.13	0.01	0.00
	A3 Age	1, <18; 2, 18-24; 3, 25-34; 4, 35-44; 5, 45-54; 6, 55-64; 7, >64	-0.06	0.08	0.10	-0.01	-0.02
	A4 Income	1, <1,000; 2, 1,000-2,000; 3, 2,001-3,000; 4, 3,001-4,000; 5, 4,001-5,000; 6 >5,000 yuan	0.38 **	0.12 **	0.21 **	0.30 **	0.26 **
	A5 Education	1, primary; 2, secondary; 3, higher	0.20 **	0.06	0.07	0.04	0.05
	A6 Occupation	1, farmers; 2, workers; 3, soldiers; 4, teachers; 5, students; 6, unemployment persons	0.22 **	0.27 **	0.19 **	0.13	0.19 **
	A7 Aim of coming	1, shopping; 2, waiting for someone; 3, passing by; 4, visiting; 5, others	0.15 **	0.19 **	0.13 *	0.18 **	0.17 **
	A8 Frequency of coming	1, once in two or three days; 2, once a week; 3, once a month; 4, once a year; 5, the first time	-0.29 **	-0.37 **	-0.20 **	-0.18 **	-0.30 **

Continued

	Variables	Measures of the attributes	Correlation or mean difference				
			Shitoudao	Railway station	Qiulin	Huizhan	Lesong
Objective variables	A9 Seasons	1, winter; 2, autumn; 3, spring; 4, summer	0.08	0.18	0.15	0.17	0.20**
	A10 Visit time	1, morning (9:00-11:59); 2, midday (12:00-14:59); 3, afternoon (15:00-18:00)	0.04	0.02	0.02	0.03	0.05
	A11 Stay time	1, less than an hour; 2, 1 - 2 hours; 3, 2 - 3 hours; 4, 3 - 4 hours; 5, more than 4 hours	-0.16**	-0.32**	-0.08	-0.14**	-0.33**
	A12 Partners	1, with partners; 2, without partners	0.03	-0.31**	0.08	0.10*	-0.04
	A13 Density of people	Number of people per m^2	0.27**	0.30**	0.34**	0.36**	0.38**
	A14 Air temperature	°C	-0.02	0.01	-0.11**	-0.01	-0.06
	A15 Relative humidity	%	-0.10	-0.29**	0.30**	0.26**	-0.33**
	A16 Horizontal luminance	lx	0.02	-0.07	-0.11**	-0.10*	0.23**
Subjective variables	B1 Reverberation evaluation	Scales from 1 to 5, with 1 as very low reverberation and 5 as very high reverberation	-0.03	0.25**	0.30**	0.03	-0.09
	B2 Acoustic comfort evaluation	Scales from 1 to 5, with 1 as very uncomfortable and 5 as very comfortable	-0.22**	-0.29**	-0.21**	-0.13**	-0.15**
	B3 Humidity evaluation	Scales from 1 to 5, with 1 as very uncomfortable and 5 as very comfortable	-0.30**	-0.33**	-0.32**	-0.08	-0.24**
	B4 Brightness evaluation	Scales from 1 to 5, with 1 as very uncomfortable and 5 as very comfortable	-0.28**	-0.29**	-0.33**	-0.22**	-0.16**
	B5 View assesment	Scales from 1 to 5, with 1 as very uncomfortable and 5 as very comfortable	-0.36**	-0.17**	-0.37**	-0.07	-0.12*

Note: ** indicates $p < 0.01$, and * indicates $p < 0.05$.

1.3.2.3　Training and testing of ANN models

In the training process the weightings are modified, such that it correctly reproduces the importance of each variable in the system[30]. However, training the network successfully requires many options and training experience[31]. In this section the data from Qiulin underground shopping street are used as a test case to explore the training and testing processes, the input variables are A1, A4, A6, A7, A8, A13, A14, A15, A16, B1, B2, B3, B4, B5 and the output variable is subjective loudness. In the model analysis process (see section 1 to 3), 80% of the data was used for training and 20% for testing because of the relatively small sample of database. In the model evaluation process (see section 4), 70% of the data was used for training, 10% for validation, and 20% for testing.

1. Layers and nodes

Previous studies demonstrated that an ANN model with three or four layers (one or two hidden layers) can be used to solve many problems[32, 33]. In this section, the number of hidden nodes with three or four layers is discussed to verify the effect of hidden layers and nodes on ANN models.

A model with three layers (one hidden layer) was selected, and hidden nodes ranged from 5 to 35. The iterations were set at 500 times for training. The training error, test error, and correlation coefficient for the test with different hidden nodes are shown in Table 1.14. The correlation coefficient for the test set increased from 0.55 to 0.68, when the hidden nodes were increased from 5 to 35. However, the correlation coefficient for the test reduced from 0.64 to 0.60, when the hidden nodes increased from 30 to 35. In other words, these results suggest the use of one hidden layer with 25 hidden nodes.

In addition, a model with four layers (two hidden layers) was examined. The number of hidden nodes in the first hidden layer ranged from 5 to 35, whereas the number of hidden nodes in the second hidden layer was kept as 5. The training sample, prediction sample, and iterations are all the same as above.

Table 1.14 Performance of models with different hidden layers and nodes

Layers	Topology	RMSE		Correlation coefficient	
		Training	Test	Training	Test
Three layers	14-5-1	0.03	0.08	0.62	0.55
	14-10-1	0.03	0.08	0.64	0.56
	14-15-1	0.02	0.07	0.65	0.56
	14-20-1	0.02	0.07	0.68	0.60
	14-25-1	0.02	0.05	0.72	0.68
	14-30-1	0.02	0.06	0.69	0.64
	14-35-1	0.02	0.07	0.65	0.60
Four layers	14-5-5-1	0.14	0.22	0.40	0.33
	14-10-5-1	0.09	0.18	0.47	0.45
	14-15-5-1	0.04	0.11	0.56	0.50
	14-20-5-1	0.04	0.11	0.54	0.51
	14-25-5-1	0.04	0.10	0.64	0.58
	14-30-5-1	0.04	0.10	0.62	0.57
	14-35-5-1	0.04	0.11	0.61	0.55

Overall, the performance of the model with three layers was better than that with four layers. It is also noted that the calculation time of the model with four layers increased by 3-5 times.

In other models, similar method and procedure was used to ensure that the best number of layers and nodes are derived.

2. Iterations

In ANN models, the number of iterations depends on the condition of data and is unknown until the whole process is completed[34]. Therefore, to find the appropriate number of iterations for an ANN model, a three-layered ANN model was created, where the number of iterations ranged from 100 to 3,000. The training error, testing error, and correlation coefficient with different number of iterations are shown in Table 1.15.

Table 1.15 Performance of models with different numbers of iteration

Iterations	RMSE		Correlation coefficient	
	Training	Test	Training	Test
100	0.13	0.18	0.48	0.38
200	0.06	0.10	0.60	0.52
300	0.05	0.09	0.64	0.56
400	0.03	0.08	0.75	0.66
500	0.03	0.08	0.75	0.67
600	0.02	0.07	0.78	0.68
700	0.02	0.07	0.79	0.68
800	0.01	0.05	0.82	0.69
900	0.01	0.06	0.80	0.67
1,000	0.01	0.06	0.77	0.65
2,000	0.00	0.08	0.68	0.62
3,000	0.00	0.09	0.62	0.58

The model performance is very low when the number of iterations is 100, with the correlation coefficient between outputs and desired targets only 0.38. The performance is acceptable when the number of iterations is from 200 to 300, with a correlation coefficient for the test set ranging from 0.52 to 0.56. The performance is generally good when the number of iterations is from 400 to 800, with a correlation coefficient for the test set ranging from 0.66 to 0.69. However, the correlation coefficient for the test set decreases from 0.67 to 0.58 when the number of iterations is increased from 900 to 3,000. With increasing number of iterations, the training error decreases continuously, and the testing error initially decreases but then increases when the number of iterations is over 800. This is perhaps due to the overfitting of the model, which is further discussed below in Section 4.

3. Functions

A three-layered ANN model was created for the prediction of subjective loudness with three classic MATLAB training functions: Levenberge-Marquardt (TRAINLM), one-step secant (TRAINOSS), and conjugate gradient with Powelle-Beale restart (TRAINCGB)[35-37]. The transfer functions used in ANN were pure linear function

(PURELIN) and tangent sigmoid function (TANSIG) (i.e. non-linear function). The iterations and hidden nodes were all fixed at 500 times and 25 nodes for training. The training error, test error, and correlation coefficient for the test with different training functions are shown in Table 1.16.

Table 1.16 Performance of models with different training functions

Training functions	RMSE		Correlation coefficient	
	Training	Test	Training	Test
TRAINLM	0.03	0.08	0.72	0.65
TRAINOSS	0.03	0.04	0.80	0.72
TRAINCGB	0.02	0.04	0.80	0.73

As shown in Table 1.16, the performance of the ANN model trained using TRAINCGB is better when compared with the others, with a correlation coefficient of 0.73 for the test set of the ANN model. The performance of the ANN model trained using TRAINLM is the worst in this case, with a correlation coefficient of 0.65 for the test set.

4. Testing of ANN models

The concept of overfitting during the training procedure is very important. This phenomenon happens to the network when the learning is performed for too long and the learner (network) is adjusted to a very specific random feature of the training data[38]. In the case of overfitting, the performance of a training example improves, but the performance for the testing data becomes worse[39]. To avoid overfitting, the validation data were used to check the training performance and stop the training when overfitting occurs. First, the data were randomly and equally divided into 10 groups, 8 of which were used for training, 1 for validation, and 1 for testing (prediction). Second, a three-layered ANN model with 25 nodes was created as the testing model with and without validation test. The iterations were all fixed at 800 times, and training function was selected as TRAINCGB (Table 1.17). The results demonstrated that the testing error of the model with validation is less than the testing error of the model without validation. Therefore, all the final models were tested at this step.

Table 1.17 Performance of models with validation test

Group	RMSE			Correlation coefficient		
Without validation	Training	Test		Training	Test	
	0.01	0.04		0.82	0.70	

Group	RMSE			Correlation coefficient		
With validation	Training	Validation	Test	Training	Validation	Test
	0.02	0.04	0.03	0.80	0.70	0.73

1.3.3 Results and discussion

Kang and Yu developed individual models, general models, and group models for the prediction of soundscapes in urban squares[19-21]. The current study therefore also developed these three kinds of model to test their performance in predicting subjective loudness in underground shopping streets.

A general model was first developed in terms of general circumstance covering various underground shopping streets, namely all of the five case study sites. To compare the effectiveness of input variables, two general models were explored, namely the G1 model and the G2 model. The inputs in the G1 model included all of the 21 variables that showed a significant correlation with subjective loudness in at least one underground shopping street in Table 1.13. The inputs to the G2 model included all the variables that showed a significant correlation with subjective loudness in more than three underground shopping streets.

The individual models were then developed in terms of prediction of subjective loudness in the different underground shopping streets, which are referred to below as the I1-I5 model. The input variables of every underground shopping street were selected individually to build the individual models.

Group models were also developed for the underground shopping streets considering those with similar characteristics. Previous research results revealed that subjective loudness in underground shopping streets can be affected by different space types or users' income[40]. Thus, the group models were classified according to space types and users' income. In the models classified by space types, the samples were divided into two groups: street type or square type. The former includes Shitoudao, Qiulin, and Lesong, whereas the latter includes the Railway station and Huizhan. The models were named S1 Model and S2 Model, respectively. In the models classified by users' income, the samples were divided into three groups based on their average income per month in Harbin: lower income (< 1,000 yuan per month), medium income (1,000-3,000 yuan per month), and higher income (> 3,000 yuan per month). The models were named U1, U2, and U3 Models.

The model parameters are given in Table 1.18. After several trials, the final optimal networks and their prediction results are shown in Table 1.19.

Table 1.18 Construction of ANN models

Models	Input variables	Output variables	Sample size
G1	A1, A2, A3, A4, A5, A6, A7, A8, A9, A10, A11, A12, A13, A14, A15, A16, B1, B2, B3, B4, B5	Subjective Loudness	1244
G2	A1, A4, A6, A7, A8, A11, A13, A15, A16, B2, B3, B4, B5		1918
I1	A1, A4, A5, A6, A7, A8, A11, A13, B2, B3, B4, B5		414
I2	A1, A2, A4, A6, A7, A8, A11, A12, A13, A15, B1, B2, B3, B4, B5		257
I3	A1, A4, A6, A7, A8, A13, A14, A15, A16, B1, B2, B3, B4, B5		307
I4	A1, A4, A6, A7, A8, A11, A12, A13, A15, A16, B2, B4		500
I5	A1, A4, A6, A7, A8, A9, A11, A13, A15, A16, B2, B3, B4, B5		365
S1	A1, A4, A6, A7, A8, A11, A13, A15, A16, B2, B3, B4, B5		1135
S2	A1, A4, A6, A7, A8, A11, A12, A13, A15, B2, B4		757
U1	A1, A4, A6, A7, A8, A11, A13, A15, A16, B2, B3, B4, B5		202
U2			1251
U3			366

Table 1.19 Final optimal networks

Model	Topology	Iterations	Training function	RMSE			Correlation coefficient (Test)
				Training	Validation	Test	
G1	21-35-1	1,000	TRAINCGB	0.09	0.13	0.10	0.44
G2	13-30-1	1,000	TRAINCGB	0.04	0.07	0.06	0.61
I1	12-30-1	800	TRAINCGB	0.02	0.04	0.03	0.74
I2	15-25-1	800	TRAINCGB	0.02	0.05	0.03	0.72
I3	14-25-1	800	TRAINCGB	0.02	0.05	0.03	0.74
I4	12-20-1	800	TRAINCGB	0.03	0.06	0.05	0.73
I5	14-25-1	800	TRAINCGB	0.04	0.07	0.06	0.67
S1	13-25-1	800	TRAINCGB	0.04	0.06	0.06	0.69
S2	11-25-1	800	TRAINCGB	0.03	0.06	0.05	0.70
U1	13-25-1	800	TRAINCGB	0.03	0.06	0.04	0.71
U2	13-20-1	800	TRAINCGB	0.03	0.07	0.05	0.71
U3	13-30-1	800	TRAINCGB	0.06	0.10	0.08	0.67

A comparison of the two general models revealed that the performance of the G1 model produced a correlation coefficient of only 0.44, which is not acceptable. A possible reason for this poor prediction result is that some input variables, such as gender, have a significant correlation with subjective loudness only in one underground shopping street. The performance of the G2 model is much better than that of the G1 model, with a correlation coefficient of 0.61 for the test set, which is considered acceptable. This result indicates that a model may be developed to predict subjective loudness for all underground shopping streets, but getting good prediction results is challenging.

The performances of individual models for subjective loudness for all the case study sites were good, with the correlation coefficient for the test set over 0.70 in four underground shopping streets. The performance of individual models in the Lesong underground shopping street is slightly lower compared with those of the other models, where the correlation coefficient for the test set of the Lesong is 0.67. This result may be due to the lower correlation between the output variable (subjective loudness) and the input variables (other variables) in Lesong underground shopping street in contrast to that in the other underground shopping streets. Overall, the successful prediction of the individual models suggests that this kind of model is useful for any special underground shopping street.

The group models for different space types performed well, with a correlation coefficient of 0.69 and 0.71 for the test set, respectively. The group models for different users' income also performed well, with a correlation coefficients of 0.71, 0.71 and 0.67 for the test set, respectively. Overall, the performance of the group models was better than the general model and slightly lower than that of the individual models. This result suggests that classifying underground shopping streets according to space type and users' income is appropriate in predicting subjective loudness.

1.3.4 Conclusions

In this study, ANN-based general models, individual models and group models were developed for the prediction of subjective loudness in underground shopping streets. Of these three models, general models have the best generality, but their performance is slightly lower than that of individual models and group models, considering that the correlation coefficient for the test of general model 2 is only 0.61. Individual models perform well, with correlation coefficients for the test set of higher than 0.70 in general. However, they have limitations because they can only be used in a given underground shopping street. Compared with the other two kinds of models, group models are the most useful and also have good prediction performance. The development of ANN models for the prediction of subjective loudness provides a new design tool for the improvement of underground shopping streets.

Once the models have been developed and adjusted, it can be concluded that in order to have a good prediction and leading to creating quiet sound environments, the perfect set of input variables for the shopping streets with street shape includes SPL, income, occupation, aim of coming, frequency of coming, stay time, density of people, relative humidity, horizontal luminance, acoustic comfort, humidity evaluation, brightness, and view; while for the shopping streets with square shape the perfect set of input variables also include partners, but exclude horizontal luminance and view.

References

[1] DING J H. A research on fire safety evacuation design in underground shopping streets [D]. Harbin: Harbin Institute of Technology, 2008: 85-86.

[2] WANG W Q. Plan and design in underground space of urban [M]. Nanjing: Southeast University Press, 2000.

[3] RESEARCH TEAMS OF CHINESE ACADEMY OF ENGINEERING (CAE). Development and utilization of underground space in urban of China [M]. Beijing: China Architecture & Building Press, 2001: 346.

[4] TONG L X. Planning and design in underground shopping streets [M]. Beijing: China Architecture & Building Press, 1996: 57.

[5] SCHAFER R M. The Soundscape: Our sonic environment and the tuning of the world [M]. New York: Destiny Books, 1994.

[6] SMYRNOVA Y, KANG J. Determination of perceptual auditory attributes for the auralization of urban soundscapes [J]. Noise Control Engineering Journal, 2010, 58 (5): 508-523.

[7] BROWN A L, KANG J, GJESTLAND T. Towards standardising methods in soundscape preference assessment [J]. Applied Acoustics, 2011, 72 (6): 387-392.

[8] KANG J, ZHANG M. Semantic differential analysis of the soundscape in urban open public spaces [J], Building and Environment, 2010, 45 (1): 150-157.

[9] YANG W, KANG J. Soundscape and sound preferences in urban squares [J]. Journal of Urban Design, 2005, 10 (1): 69-88.

[10] YANG W, KANG J. Acoustic comfort evaluation in urban open public spaces [J]. Applied Acoustics, 2005, 66 (2): 211-229.

[11] NILSSON M E, BERGLUND B. Soundscape quality in suburban green areas and city parks [J]. Acta Acustica united with Acustica, 2006, 92 (6): 903-911.

[12] BARLOW P, TEASDALE G M, JENNETT B, et al. Computer assisted prediction of outcome of severely head-injured patients [J]. Journal of Microcomputer Applications, 1984, 7 (3-4): 271-277.

[13] LANG E W, PITTS L H, DAMRON S L, et al. Outcome after severe head injury: an analysis of prediction based upon comparison of neural network versus logistic regression analysis [J]. Neurological research, 1997, 19 (3): 274-280.

[14] FABBRI A, SERVADEI F, MARCHESINI G, et al. Early predictors of unfavourable outcome in subjects with moderate head injury in the emergency department [J]. Journal of neurology, neurosurgery, and psychiatry, 2008, 79 (5), 567-573.

[15] SMITS M, et al. Outcome after complicated minor head injury [J]. American journal of neuroradiology, 2008, 29 (3): 506-513.

[16] Pang B C, et al. Hybrid outcome prediction model for severe traumatic brain injury [J]. Journal of Neurotrauma, 2007, 24 (1): 136-146.

[17] SCHULTE-FORTKAMP B. Soundscape analysis in a residential area: An evaluation of noise and people's mind [J]. Acta Acustica united with Acustica, 2006, 92 (6): 875-880.

[18] NANDA S K, TRIPATHY D P, PATRA S K. A soft computing system for opencast mining machineries noise prediction [J]. Noise Control Engineering Journal, 2011, 59 (5), 432-446.

[19] KANG J. Urban Sound Environment [M]. London: Taylor and Francis, 2004: 73-76.

[20] YU L. Soundscape Evaluation and ANN Modeling in Urban Open Spaces [D]. Sheffield: University of Sheffield, 2009: 81-86.

[21] YU L, KANG J. Modeling subjective evaluation of soundscape quality in urban open spaces: An artificial neural network approach [J]. Journal of the Acoustical Society of America, 2009, 126 (3): 1163-1174.

[22] EMBRECHTS M J. Neural network for date mining [M] // CHEN P, FERNANDEZ B R, GOSH J. Intelligent engineering systems through artificial neural networks. New York: ASME Press, 1995.

[23] WEI H K. Theories and methods in ANN design [M]. Beijing: National Defense Industry Press, 2005: 33.

[24] MENG Q, JIN H, KANG J. Soundscape evaluation in different types of underground shopping streets [J]. Journal of Harbin Institute of Technology, 2011, 18 (suppl. 2): 231-234.

[25] MENG Q, JIN H, KANG J. The influence of users' behavioral characteristics on the evaluation of subjective loudness and acoustic comfort in underground shopping streets [J]. Huazhong Architecture, 2010: 28 (5), 90-92.

[26] BROWN A L. Soundscapes and environmental noise management [J]. Noise Control Engineering Journal, 2010, 58 (5): 493-500.

[27] MENG Q, JIN H, KANG J. Approach on the survey method for sound environment in underground spaces [J]. Advanced Materials Research, 2012, 457-458: 229-232.

[28] MENG Q, JIN H, KANG J. On the soundscape survey method in underground shopping streets [G]. INTER-NOISE and NOISE-CON Congress and Conference Proceedings,

InterNoise08, Shanghai CHINA: 1306-1311 (6).

[29] MENG Q, KANG J, JIN H. Field study on the influence of users' social qualities on the evaluation of subjective loudness and acoustic comfort in underground shopping streets [J]. Applied Acoustics, 2010, 29 (5): 371-381.

[30] HAN R F, LI J. Research of self-adaptation mixed genetic algorithm based on BP operator [J]. Computer Engineering and Design, 2007, 28: 651-652.

[31] VERMA L S, SHROTRIYA A K, SINGH R, et al. Prediction and measurement of effective thermal conductivity of three-phase systems [J]. Journal of Physics D: Applied Physics, 1991, 24: 1515-1526.

[32] JIAO L C. Calculation in Neural Network [M]. Xi'an: Xi Dian University Press, 1992: 25-36.

[33] ZHANG L M. Models and Applications of Artificial Neural Network [M]. ShanHai: Fu Dan University Press, 1993: 13-47.

[34] LI J, ZENG H L. Function optimization on mixed intelligence learning algorithm [J]. Computer Measurement & Control, 2007, 15: 1067-1068.

[35] GE Z X, SUN Z Q. Theories in Artificial Neural Networks and Test in Matlab R2007 [M]. Beijing: Electronics industry Press, 2008, 3: 102.

[36] HAN M, DING J. Improvement of BP algorithm based on cross-validation method and its implementation [J]. Computer Engineering and Design, 2008, 29: 3738-3739.

[37] ABBODA M F, CHENGB K Y, CUIC X R, et al. Ensembled neural networks for brain death prediction for patients with severe head injury [J]. Biomedical Signal Processing and Control, 2011, 6: 414-421.

[38] LIU J G, JOLLY R A, SMITH A T, et al. A New Algorithm to Reduce Overfitting for Genomic Biomarker Discovery [J]. Predictive Power Estimation Algorithm, 2011, 6: e24233.

[39] NASON G P. Wavelet regression by cross-validation [J]. Journal of the Royal Statistical Society, 1996: 58, 463-479.

[40] MENG Q. Research and Prediction on Soundscape in Underground Shopping Streets [D]. Harbin: Harbin Institute of Technology, 2010: 85-86.

Chapter 2

Sound Environment and
Acoustic Perception
in Dining Spaces

2.1 Effects of typical dining styles on conversation behaviours and acoustic perception in restaurants

2.1.1 Introduction

Restaurants have expanded their significance beyond dining alone and become places of emotional communication, family gatherings, and commercial negotiations. Field research by Heung and Gu[1] found that customers' choice of restaurants is not limited to the consideration of food factors any longer; the restaurant environment, particularly the sound environment, considerably affects diners' evaluation of their comfort and the overall dining experience.

The sound environment and users' acoustic perception in restaurants are a common focus of research, since the evaluation of meals and the income of restaurants can be affected strongly by sound factors[2]. Various studies have examined acoustic problems in dining spaces, including those related to noise control, speech intelligibility, and acoustic comfort[3-5]. Regarding noise control, Kang and Lok[6] found that the background noise level in restaurants is generally 80-90 dB(A), while the ideal noise level is 70-75 dB(A)[7]. The acoustic environment in restaurants can be substantially affected by equipment noise, including lampblack machines and fans, and using any type of stone material for sound absorption in restaurants is not optimal[8,9]. To examine speech intelligibility in restaurants, Kang[10] used a radiosity-based computer model to establish a mathematical model which revealed that increasing boundary absorption typically increases the speech transmission index (STI) by 0.2-0.4. With certain reverberation times, unintelligible speech sounds are expected to act as masking sounds, so that communication among diners around the same table will not be disturbed by the noise of diners at neighbouring tables.

While the sound level, threshold of background noise that sheltered the noise interference of diners at neighbouring tables and guaranteed their speech articulation was found to be relatively narrow, at 69-71 dB(A)[11-13]. In terms of acoustic comfort, Leccese et al. [14] proposed a simplified analytical model to evaluate the acoustic conditions required to ensure the intelligibility of conversations in restaurant dining rooms, and found that the "cocktail party effect" significantly affected the level of comfortable acoustic conditions. Another study, on two typical large dining spaces, found that background music, other diners' speech sounds, and impact sound from tableware had the dominant impacts on acoustic comfort evaluations by diners[7].

The conversation of diners is one of the main behaviours influencing the sound environment and diners' acoustic perceptions in restaurants. Ariffin et al. [15] studied the influence of environmental factors including colour, lighting, design, and layout on the conversation behaviour of diners. The Lombard effect or Lombard reflex is the involuntary tendency of speakers to increase their vocal effort when speaking amid loud noise, to enhance the audibility of their voice[16]. Field research has found that the listener may follow conversation of interest despite many concurrent sources of sound[17].

The indoor and outdoor sound environment and users' acoustic perception can also be affected by crowd density[18, 19], since a crowd is a special sound source in that it gives rise to certain sound absorption effects[20-22]. Studies have found that the sound environment in commercial pedestrian streets and underground shopping streets has undergone many changes, and acoustic comfort—as a key evaluation index of acoustic perception—varies substantially with crowd density [23]. Meng and Kang put forward a crowd acoustic model applicable to large spaces and applied a method of equivalent sound source calculation along with a simplified method for crowd sound sources [24]; in a separate study, Nie and Kang also analysed the relationship between crowd density and sound pressure level and between the number of persons present and the number of persons conversing [25]. However, few studies have considered the influence of the crowd factor on conversation behaviours and acoustic perception in restaurants.

Background music, which is a common sound source inrestaurants, may also affect the sound environment and users' acoustic perceptions. Previous studies have

Chapter 2　Sound Environment and Acoustic Perception in Dining Spaces

indicated that the acoustic comfort of customers in commercial spaces is higher with than without background music[26]. In restaurants, previous studies have been confined to the influence of background music on eating behaviour, dining rate, meal volume, and sensitivity to food, without taking the influence of background music on conversation behaviour into account [27-30].

Thus, the aim of this study is to find out the effects of typical dining styles on conversation behaviours and acoustic perception in restaurants in China. First, this study examined the influence of dining styles on conversation behaviours, such as diners' frequency of conversation and frequency of speech sound. Second, the influence of dining styles on sound pressure level in restaurants was studied. Third, the influence of dining styles on acoustic comfort of diners was investigated. Three typical dining styles, including centralized, separate and dispersed styles, were compared. Crowd density and restaurants with and without background music were considered in this study, as two factors which may affect conversation behaviours and acoustic perception in restaurants.

Some key terms used in this paper are defined/explained below.

(1) Dining styles. Based on the analysis of relevant studies[31-34], this study divides dining styles into three categories: centralised, separate and dispersed, as shown in Fig. 2.1. The centralised style refers to diners sharing a dish, such as a hot pot; the separate style means that diners do not share dishes with others but eat their own food; while in the dispersed style, diners share many dishes, which is common in family gatherings (see Fig. 2.1). Previous studies have shown that these three dining styles are common not only in China but also in other countries in Europe and Asia [35].

(2) Conversion behaviours. This study considers two kinds of conversion behaviour, namely the frequency of conversation and the frequency of speech sound. The former indicates the proportion of the time of a diner having conversation with any diner at the same table. The latter indicates the proportion of the time of a diner heard surrounding speech.

(3) Sound pressure level. It is a logarithmic measure of the effective pressure of a sound relative to a reference value, and the unit of sound pressure level is dB [11].

(4) Acoustic comfort. It is the subjective evaluation of a diner on the dining environment, with a five-point scale in this study: 1, very uncomfortable; 2, uncomfortable; 3, neither comfortable nor uncomfortable; 4, comfortable; 5, very comfortable[26].

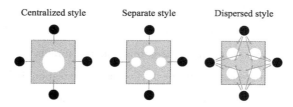

Fig. 2.1　Three dining styles: centralized, separate and dispersed

2.1.2　Methodology

2.1.2.1　Survey site

The selection of case restaurants for acoustic studies is important, since many factors may affect sound environment and acoustic perception in restaurants[7, 10, 36, 37]. A study has indicated that compared with general restaurants, fast-food restaurant may be less noisy, since dining periods of less than half an hour will involve less conversation than those of more than 1 hour[36]. Kang pointed out that the different geometry of restaurants may change their reverberation time (RT)[10]. Previous studies have also pointed out that sound environment can change with the interior layout of the restaurant; for instance, a restaurant with the kitchen inside usually has higher sound level than one with the kitchen adjacent[7]. Some recent studies have indicated that table shape can also affect users' acoustic perceptions in indoor spaces; for example, when the length of the table is 5 times the width [37].

Thus, based on a preliminary study, 523 Chinese restaurants were surveyed to find out their typical features, considering dining style, geometry, layout and the social and behaviours characteristics of diners[38]. Each of the above mentioned three dining styles in this study was investigated, with and without background music.

Consequently, six restaurants were chosen to cover all these situations.

The restaurants with music were Hongming Hot Pot (HHP), with a centralized style, Alpine Buffet (AB), with a separate style, and Bee Kitchen (BK), with a dispersed style. The restaurants without music were Sichuanren People Hotpot (SCR), with a centralized style, Hawaiian Pizza (HWP), with a separate style, and Chuanrenbaiwei (CRBW), with a dispersed style. To avoid unusual influences of space and scale on sound distribution[39], the proportions of these six restaurants (length, width and height) were within a ratio of 1 : 3, in order to avoid extremely non-diffuse sound fields[40]. Some details of the case sites, such as restaurant style, capacity, geometry and indoor photographs, are shown in Table 2.1. As previous studies have indicated that the evaluation of acoustic environment can be influenced by reverberation time[41], the measured unoccupied RT for each of the six restaurants is also given in Table 2.1. It can be seen that the difference in RT T30 is less than 0.1s; therefore, the influence of reverberation time was not taken into account in this study [42-44]. Rindel [45] found that the typical number of persons at a table was 2-6, which was so in the cases of the restaurants listed above. To avoid the influence of the signal-to-noise ratio of the restaurant's public-address (PA) system on behaviour patterns and acoustic perception [45], the same acoustic system and background music were used in all the selected restaurants (that had background music). The type and tempo of music used [27, 46] are shown in Table 2.1.

Table 2.1 The basic information about six typical restaurants

Restaurants	HHP	SCR	AB	HWP	BK	CRBW
Dining style	Centralized	Centralized	Separate	Separate	Dispersed	Dispersed
Volume	$910m^3$	$910m^3$	$1208m^3$	$1190m^3$	$942m^3$	$1173m^3$
Geometry (length/width)	Rectangle 13m/20m	Rectangle 14m/19m	Rectangle 18m/19m	Rectangle 17m/20m	Rectangle 14m/19m	Rectangle 17m/18m
Shape of table (length/width)	Rectangle 1.5m/1m	Rectangle 1.5m/1m	Quadrate 1m/1m	Rectangle 1.2m/1m	Rectangle 1.5m/1m	Quadrate 1m/1m
Photograph						

Continued

Restaurants		HHP	SCR	AB	HWP	BK	CRBW
Interior materials and sound absorption coefficient	Ceilings	Gypsum $\alpha = 0.3$	Gypsum $\alpha = 0.3$	Gypsum $\alpha = 0.3$	Gypsum $\alpha = 0.3$	Gypsum $\alpha = 0.3$	Gypsum $\alpha = 0.3$
	Walls	Ceramic $\alpha = 0.02$	Plaster $\alpha = 0.01$	Wood $\alpha = 0.03$	Marble $\alpha = 0.01$	Ceramic $\alpha = 0.02$	Plaster $\alpha = 0.01$
	Floors	Marble $\alpha = 0.01$	Ceramic $\alpha = 0.02$	Ceramic $\alpha = 0.02$	Marble $\alpha = 0.01$	Ceramic $\alpha = 0.02$	Marble $\alpha = 0.01$
Reverberation time		1.57s	1.58s	1.61s	1.57s	1.58s	1.58s
Music		With music	Without music	With music	Without music	With music	Without music
Music style		pop		pop		pop	
Music tempo		95-100bpm		95-100bpm		95-100bpm	
Price level CNY/USD*		62 yuan 8.9 dollars	80 yuan 10.1 dollars	63 yuan 9.1 dollars	63 yuan 9.1 dollars	56 yuan 8.1 dollars	54 yuan 7.8 dollars
Age segment of diners		17-44	22-40	18-42	18-43	20-40	15-46

Note: * According to Bank of China, the average exchange rate between China yuan and the US dollar in 2016 is 6.9125.

2.1.2.2 Crowd density measurement

Previous studies have shown that crowd density was a key influence on the acoustic environment and acoustic perception in open and indoor urban spaces[21]. Given this, we might expect conversation behaviour in restaurants as well to be influenced by crowd density; thus, this study also measured crowd density with each of the dining styles. Measurements were performed every half hour between the hours of 10:00 and 22:00 to cover variations in occupancy rate over time, as shown in Fig. 2.2[47]. In order to reduce measurement error, volunteers were asked to measure the number of diners at the same time, each covering 2-3 tables. Further, cameras recorded the scene, and numbers were confirmed through video playback in the laboratory[21]. At the end of each measurement, the numbers collected by the volunteers were weighted to get the total number of diners. Finally, crowd density was calculated as the total number of diners divided by the area of the restaurant [48].

Chapter 2　Sound Environment and Acoustic Perception in Dining Spaces

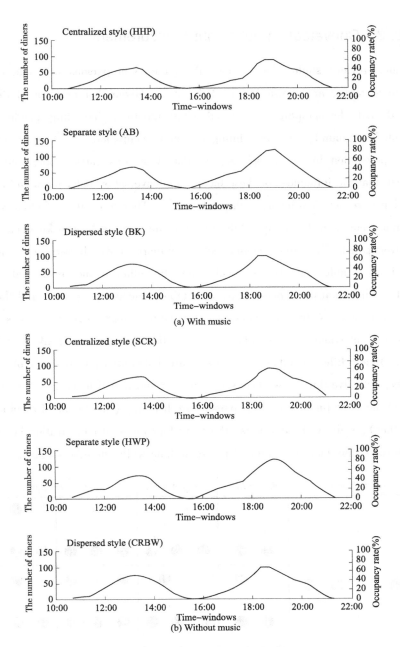

Fig. 2. 2　Trend and variation in number of diners and occupancy rate with time-windows

2.1.2.3 Conversation behaviour measurement

Some previous studies have shown the patterns of persons who talk and are talked to at table, are shown in Fig. 2.3 [49, 50]. Given that previous studies had indicated that demographic and social characteristics, including gender, age, education background, income, dining out and occupation [7, 38], may influence acoustic perception in indoor spaces, a pilot study was carried out, and it was confirmed that the influences of these factors on conversation behaviours and acoustic perception were not significant, with $p > 0.1$ [50]. Therefore, these factors were not taken into account in this study. Only diners at the same table who knew each other were investigated. The above-mentioned preliminary study showed that when two diners sit at a table, one is the speaker and the other is the listener at a given moment, while when three persons are at a table, one is the speaker and the other two are listeners. With four persons at a table, there may be two kinds of conversation behaviours: one speaker and three listeners, or two speakers talking at the same time while the other two persons are listening (one to each of them). Similarly, with five diners, we may see a speaker and four listeners, or two speakers talking at the same time while three persons are listening (one to one of them and two to the other); while with six diners, there are four kinds of conversation behaviour: one to three diners may be talking at the same time while the others listen.

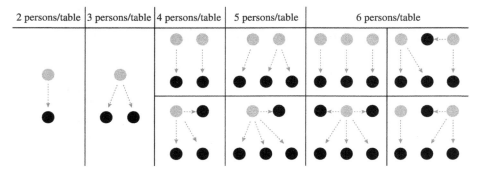

Fig. 2.3 Conversation patterns

Note: Grey circle indicates persons who talk, black circle indicates persons who listen.

Both frequency of conversation and frequency of speech sound were measured in this study. Frequency of conversation was measured across three periods into which the dining process was divided: waiting, eating and after-eating. The waiting period begins when diners are seated and ends when they start eating; the eating period lasts from when they start eating to when 75% or more of them have stopped eating; and the after-eating period starts at the end of the meal and ends when the diners leave their seats. For measurements, an HD video camera was used to record the dining process. The video recorded the total time of each period, and the speaking time of each diner in each dining period was measured using a stopwatch. Speaking time was estimated from when one diner started speaking to when he/she stopped. When two or more diners spoke at the same time, speaking time was only measured once. The duration of each diner's speech was weighted and divided by the total time to obtain the frequency of conversation. In every restaurant, after each measurement of crowd density 10-15 tables were randomly chosen and the diners were asked to sit for an interview. These interviewees were asked to describe the 1-3 most salient sounds [32] that they had heard and to categorize them as speech, background music (if there was any), or other sounds[51, 52]; then, the frequency of different sounds was calculated and divided by the total frequency of all sounds to get proportions for each.

2.1.2.4 Sound pressure level measurement

Previous studies have suggested that different crowd aggregation states and behaviour patterns influence the sound environment and acoustic perception of users in open and indoor spaces, and the sound environment can in turn influence their acoustic perception[38]. Therefore, the level of sound pressure in every restaurant was measured, with the following methods, immediately after each measurement of crowd density. The sound pressure level meter was set to slow-mode and A-weight, and a reading for instantaneous data was taken every 10s. The probe of the sound level meter was positioned 1m away from walls and other main reflectors and 1.2-1.5m off the ground[53, 54]. A total of 5 minutes of data were obtained at each measurement position, and the corresponding A-weight equivalent sound pressure level was derived. In order to avoid measurement error, each measurement in each restaurant was taken from at least five random points, with a distance between each point of at least 3m[55]. In order to avoid the impact of persons speaking on the measurement, there were no persons talking within 3m of the scope[55]. The A-weight sound pressure levels measured at each point were averaged[56].

2.1.2.5 Acoustic comfort survey

Acoustic comfort is a key evaluation index for the soundscape of open and indoor spaces[7, 21]. Thus, this study examined the influence of different dining styles on the evaluation of diners' acoustic comfort. After the measurement of crowd density and sound pressure level, some diners were immediately extracted and invited to take a questionnaire survey (questions and scales are in Table 2.2). The questions covered diners' social characteristics, and acoustic comfort[57]. The interviewers were instructed to explain questions and ensure that interviewees understood them. The interviewees, who were diners chosen randomly from the case sites, were asked to assess the acoustic comfort of the restaurant. On acoustic comfort, interviewees answered on the following five-point Likert-type scale: 1, very uncomfortable; 2, uncomfortable; 3, neither comfortable nor uncomfortable; 4, comfortable; 5, very comfortable[57]. Before the formal investigation, the validity and reliability of the questionnaire were tested[58-60]. As previous studies indicated that users may not be able to evaluate an acoustic environment accurately until around 15min after they have entered it[61], the interviews were carried out about 20-30 min after diners entered the restaurant. Previous studies have also indicated that an interview of more than 5 minutes may decrease the reliability of investigation[61-62], so the questionnaires in this study were all delivered and finished within 2-3 minutes.

To ensure the representativeness of the results, a survey on the social characteristics of diners, including age, gender, income and education, was also done in all six restaurants before the formal investigation[38]; there was no significant difference found between these social characteristics of dinners in the preliminary survey and in the formal investigation (mean difference 0.01-0.04 with $p > 0.1$). It was shown that the results obtained from the six restaurants are typical.

Table 2.2 Questionnaire questions and scales

Questions	Scale
Gender	1, male; 2, female
Age	1, <18; 2, 18-24; 3, 25-34; 4, 35-44; 5, 45-54; 6, 55-64; 7, >64
Income (yuan)	1, <1,000; 2, 1,000-2,000; 3, 2,001-3,000; 4, 3,001-4,000; 5, 4,001-5,000; 6, >5,000
Education level	1, primary; 2, secondary; 3, higher
Occupation	1, farmer; 2, industrial worker; 3, soldier; 4, teacher; 5, student; 6, unemployed person
Visit time	1, morning (9:00-11:59); 2, midday (12:00-14:59); 3, afternoon (15:00-18:00)
Stay time	1, less than one hour; 2, 1-2 hours; 3, more than 2 hours
Acoustic comfort	scale from 1 to 5, with 1 as very uncomfortable and 5 as very comfortable

2.1.2.6 Statistics and analysis

SPSS 15.0[63] was used to establish a database for all the subjective and objective measurements. The Pearson correlation was used to calculate the relationships between crowd density and sound pressure level and between crowd density and diners' comfort evaluation. The linear and nonlinear regression analyses were used to establish the regression equations of crowd density and sound pressure level, and crowd density and diners' comfort evaluation. The t-test at $p < 0.01$ and $p < 0.05$ was used to test sound perception with and without background music.

2.1.3 Results

2.1.3.1 Influence of dining styles on conversation behaviour

1. Frequency of conversation

The frequency of conversation of diners in the waiting, eating and after-eating periods is analysed first. The maximum difference in frequency of conversation between the three dining periods was 6.8%, which is not significant; therefore, the periods were merged to analyse the influence of dining styles on frequency of conversation, frequency of speech sound, and subjective experience of diners. Fig. 2.4 shows the influence of increasing the number of diners at each table on frequency of conversation. Restaurants with and without background music were considered.

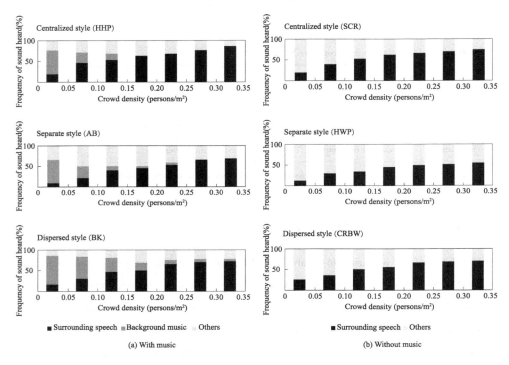

Fig. 2.4 Influence of increasing number of diners at each table on frequency of conversation in three typical dining styles

With background music. As Fig. 2.4(a) shows, in restaurants of a centralized style, frequency of conversation of diners was the highest when five persons were at each table, reaching a value of 81.5%. In the separate and dispersed styles, it was highest with six persons, with values of 72.2% and 76.0%, respectively. Average frequency of conversation was 69.8% with centralized, 65.6% with dispersed, and 58.0% with separate style. This might emerge from the characteristics of each dining style: in the centralized style, diners are seated around a brazier to share a dish, which would plausibly account for the increased frequency of conversation. In contrast, there were many dishes in the dispersed style, so that frequency of conversation was relatively moderate. Finally, in the separate style, diners use tableware individually and communications between diners are less frequent, resulting in the lowest frequency of conversation.

Without background music. As Fig. 2.4(b) shows, in the centralized, dispersed, and separate styles respectively, when six persons were at each table the frequency of conversation was 85.1%, 77.8%, and 66.7%. Similarly, the average frequency of conversation of diners in centralized dining was the highest, with a value of 66.8%, while the average frequency of conversation of diners with separate and dispersed styles was 64.6% and 54.8%, respectively.

In conclusion, four persons or more per table reduced the frequency of conversation of diners effectively, as did the separate and dispersed styles. It is also interesting to note that the average frequency of conversation of diners in restaurants with background music was higher than that in restaurants without background music, with mean differences of 3.2% in separate dining, 3.0% in centralized dining, and 1.0% in dispersed dining ($p < 0.01$). A possible reason is that when playing background music, the diners felt that their privacy improved; thus, an acoustic environment with music is more suitable to help diners chat. Another reason may be that the comfort of diners improved, so that they wanted to talk more when background music played.

2. Frequency of speech sound

In the three typical dining styles, the relationship between crowd density and the frequency of sound heard is shown in Fig. 2.5; restaurants with and without background music were considered.

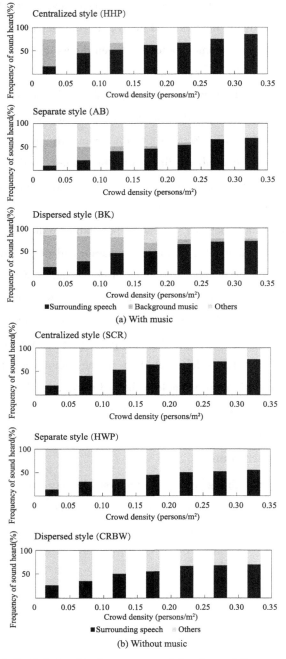

Fig. 2.5 Relationship between crowd density and frequency of sound heard in the three typical dining styles

With background music. As Fig. 2.5(a) shows, the percentage of diners who heard speech increased with increasing crowd density, which corresponds to previous results for studies in urban open spaces[21]. When the crowd density was between 0 and 0.1 persons/m^2, the frequency of speech sound in the centralized style was 30.9%, which was 16.4% higher than that in separate style and 8.4% higher than that in dispersed style. When crowd density ranged from 0.25 to 0.35 persons/m^2, the frequency of speech sound in the centralized style was 80.4%, which was 13.7% higher than with separate style and 9.2% higher than with dispersed style. With different dining styles, crowd density increased by 0.05 persons/m^2, and the average frequency of speech sound increased by 11.3% with the centralized style, 9.8% with the separate style, and 9.3% with dispersed style.

Without background music. As Fig. 2.5(b) shows, when crowd density ranged from 0 to 0.1 persons/m^2, the frequency of speech sound in the centralized style was 30.4%, which was equivalent to that in the dispersed style and 8.9% higher than in the separate style. When crowd density ranged from 0.25 to 0.35 persons/m^2, the frequency of speech sound in the centralized style reached 73.1%, which was 19.6% higher than in the separate style and 3.8% higher than in the dispersed style. With different dining styles, crowd density increased by 0.05 persons/m^2, and the average frequency of speech sound increased by 9.3% with the centralized style, 7.5% with the dispersed style, and 7.0% with the separate style.

These results indicate not only that the frequency of speech sound in the centralized style was higher than that in the other two styles, but that it also had a greater increment of the frequency of speech sound with increasing crowd density than the other two dining styles. When crowd density was lower than 0.15 persons/m^2 in all three dining styles, as crowd density increased the increment of the frequency of speech sound was higher than for 0.15 persons/m^2 and higher crowd density ranges, where privacy derived from distance disappeared and so frequency of speech sound increased only slowly with density. In addition, the average frequency of speech sound in restaurants with background music was higher than in restaurants without background music, with a difference of 5.2% in the separate style, 4.3% in the centralized style, and 1.2% in the dispersed style ($p < 0.01$), which matches the results on frequency of conversation, which showed it higher with background music than without.

2.1.3.2　Influence of dining styles on sound pressure level

As seen, diners' conversation behaviours differed between dining styles. This section discusses the influence of different dining styles on sound pressure in restaurants, considering cases with and without background music. Fig. 2.6 shows the relationship between crowd density and measured sound pressure level in these six restaurants, with the corresponding linear trend curves and coefficient of determination R^2 and with $p < 0.001$. Various regressions, including linear, quadratic, and cubic, were used to find out the best fit to show relationships between crowd density and measured sound pressure level; Fig. 2.6 (a) shows a centralized style without music as an example. It can be seen from the figure that the linear regression for this restaurant is better than the others, with $R^2 = 0.993$. Therefore, in the following analysis, this linear regression is used to explain the relationships between crowd density and measured sound pressure level.

1. With background music

As Fig. 2.6 (b) shows, when crowd density increased from 0.05 to 0.25 persons/m^2, the level of sound pressure inside the restaurant increased accordingly, by 5.9dB(A) in the separate style, 5.3dB(A) in the centralized style, and 4.1dB(A) in the dispersed style. For the same crowd density, the sound pressure level in the centralized style was higher than in the separate and dispersed styles, with values of 2.5dB(A) and 2.0dB(A), respectively. One reason for this may be the fact that in the case with background music, frequency of conversation in the centralized style was higher than in the separate style, with a difference of 11.8%, and the dispersed style, with a difference of 4.2%. In the separate and dispersed styles, conversely, under the same crowd density, in the range of 0 to 0.18 persons/m^2, mean sound pressure in the dispersed style was higher than in the separate style, while from 0.18 to 0.35 persons/m^2, it was higher in the separate style. This seems to show that when background music is played, the separate style can reduce sound pressure effectively when crowd density is less than 0.18 persons/m^2, while when crowd density exceeds 0.18 persons/m^2, the dispersed style can reduce the level of sound pressure more effectively.

Chapter 2 Sound Environment and Acoustic Perception in Dining Spaces

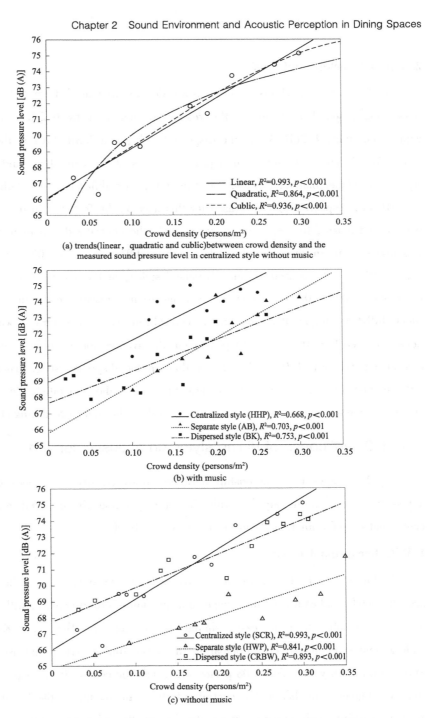

Fig. 2.6 Relationship between crowd density and measured sound pressure level in six restaurants

2. Without background music

As Fig. 2.6(c) shows, when crowd density increased from 0.05 to 0.25 persons/m^2, sound level inside the restaurant increased by 6.2dB(A) in the centralized style, 4.2dB(A) in the dispersed style, and 3.3dB(A) in the separate style. At this crowd density, sound pressure levels of the centralized and dispersed styles were higher than that of the separate style, with difference of 4.4dB(A) and 3.8dB(A), respectively. One reason for this may be the fact that in the absence of background music, frequency of conversation in the centralized and dispersed styles was higher than in the separate style, with a difference of 12.0% and 9.8%, respectively. In the centralized and dispersed styles, at the same crowd density, in the range from 0 to 0.17 persons/m^2, average sound pressure in the dispersed style was 0.8dB(A) higher than that in the centralized style. In contrast, when crowd density ranged from 0.17 to 0.35 persons/m^2 the average sound pressure level in the centralized style was 1.0dB(A) higher than in the dispersed style. These results indicate that in the absence of background music, the separate style can more effectively reduce sound pressure in restaurants than the other two dining styles.

2.1.3.3 Influence of dining styles on acoustic comfort

Fig. 2.7 shows the relationship between crowd density and acoustic comfort across the three dining styles, with the corresponding linear trend curves and coefficient of determination R^2 and significance $p < 0.001$.

1. With background music

As Fig. 2.7(a) shows, in the centralized and separate styles, the value of acoustic comfort takes a parabolic shape as a function of crowd density, with an initial increase and subsequent decrease, similar to the study by Meng and Kang[24] on underground shopping streets. Comparing these two dining styles demonstrated that when crowd density exceeded 0.08 persons/m^2, acoustic comfort in the separate dining style was higher than that in the centralized style. Interestingly, the linear trend of acoustic comfort with varying crowd density in the dispersed dining style was significantly different from the other two dining styles: with increasing crowd density,

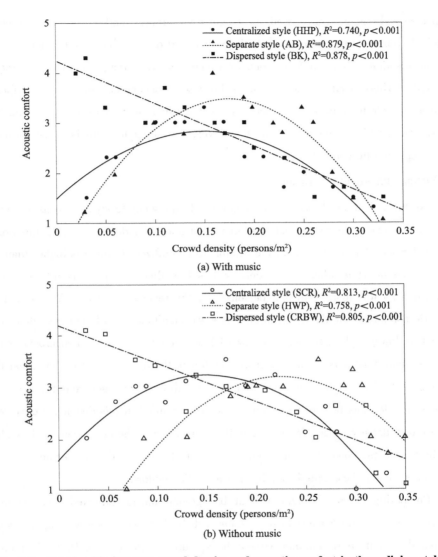

Fig. 2.7 Relationship between crowd density and acoustic comfort in three dining styles

the acoustic comfort of diners decreased. It was higher than that in the other two dining styles when crowd density ranged from 0 to 0.12 persons/m², and lower at higher densities. The reasons for these different trends may be that the centralized and separate styles are mainly used for family gatherings, and diners dining in these two styles prefer these busy establishments. In such a case, a sound environment with lower crowd density could make customers feel less cheerful. On the other hand, the dispersed style is generally used for commercial dining, and these diners prefer a relatively quiet dining environment with lower crowd density that may be more suitable for conversation (as when crowd density exceeds a given range, normal exchange is affected). These results indicate that in the case of background music, the dispersed style is suitable to achieve better acoustic comfort when crowd density is less than 0.12 persons/m^2, while when it exceeds that level, the separate style is better suited to achieving acoustic comfort.

2. Without background music

As Fig. 2.7 (b) shows, in the centralized and separate styles, acoustic comfort also took a parabolic shape that first increased and then decreased as a function of crowd density. The value of acoustic comfort in centralized dining was higher than that in the separate style when crowd density was less than 0.2 persons/m^2, but when crowd density was more than 0.2 persons/m^2, the reverse was true. Also as before, acoustic comfort in the dispersed dining style was significantly different from that in the other two dining styles, taking a downward linear trend. Overall, when crowd density was less than 0.12 persons/m^2, acoustic comfort in the dispersed style was higher than in the centralized style, showing that in the absence of background music, the dispersed styles was best suited to ensure acoustic comfort at this density. Conversely, when crowd density ranged from 0.12 to 0.2 persons/m^2, the centralized style yielded the best acoustic comfort, and when crowd density exceeded 0.2 persons/m^2, the separate style was best suited to achieving acoustic comfort.

Comparing cases with and without background music indicated that the mean difference in acoustic comfort in restaurants with background music was higher by 0.6 compared to that in restaurants without background music ($p < 0.01$) when crowd density ranged from 0 to 0.23 persons/m^2, whereas when crowd density ranged from 0.23 to 0.35 persons/m^2, the reverse was true ($p < 0.01$). Consequently, when crowd density is less than 0.23 persons/m^2, background music can be played to achieve better acoustic comfort; but when crowd density is greater than 0.23 persons/m^2, background music is not conducive to acoustic comfort.

2.1.4 Conclusions

Based on objective measurements and a subjective survey of six typical Chinese restaurants, this study examined the differences between conversation behaviour and acoustic perception of diners in three styles of restaurant, respectively featuring centralized, separate and dispersed dining. The following conclusions can be drawn from the results:

First, regarding the influence of dining styles on conversation behaviour, centralized dining will increase frequency of conversation. The presence of four or more persons at each table can also increase frequency of conversation effectively. It is interesting to note that, with the same crowd density, the frequency of conversation of diners in restaurants with background music was higher than that in restaurants without background music.

Second, as crowd density increased, sound pressure inside the restaurant increased as well. In restaurants with background music, the separate style reduced sound pressure most effectively when crowd density was less than 0.18 persons/m^2. In the absence of background music, the separate style reduced sound level most effectively across the board.

Third, regarding the influence of dining styles on acoustic comfort, in the centralized and separate styles, acoustic comfort took on a parabolic shape, while in the dispersed style it decreased linearly as crowd density increased. With background music, the dispersed style achieved better acoustic comfort when crowd density was less than 0.12 persons/m^2, while the separate style achieved better acoustic comfort when crowd density was more than 0.12 persons/m^2. In the absence of background music, when crowd density was less than 0.12 persons/m^2, the dispersed style achieved the best acoustic comfort, while when crowd density was greater than 0.2 persons/m^2, the separate style was best suited to ensure acoustic comfort.

While this study is based only on typical Chinese restaurants, a previous study has shown that the size of restaurants in Europe is usually 1/3-1/4 of that of Chinese restaurants[64]. This could lead to rather different reverberation times, which could influence the acoustic comfort of diners[10]. Moreover, another previous study pointed out that the vary price of restaurants may lead to social differences, which in turn, could influence the conversation behaviours[65]. Furthermore, some recently works have shown that the different background music styles, such as jazz, rock-and-roll and classical music may also lead to the different speed of conversation[66], and the acoustics comfort could consequently be affected. Thus, in future studies, those could be further examined.

References

[1] HEUNG V C S, GU T. Influence of restaurant atmospherics on patron satisfaction and behavioral intentions [J]. International Journal of Hospitality Management, 2012, 31 (4): 1167-1177.

[2] HARDY H C. Cocktail party acoustics [J]. The Journal of the Acoustical Society of America, 1975, 57 (s1): 535.

[3] WHITE A. The effect of the building environment on occupants: the acoustics of dining spaces [D]. Cambridge, UK: University of Cambridge, 1999.

[4] LEE P J, KIM Y H, JEON J Y, et al. Effects of apartment building facade and balcony design on the reduction of exterior noise [J]. Building and Environment, 2007, 42 (10): 3517-3528.

[5] LEE P J, JEON J Y. Evaluation of speech transmission in open public spaces affected by combined noises [J]. The Journal of the Acoustical Society of America, 2011, 130 (1): 219-227.

[6] KANG J, LOK W. Architectural acoustic environment, music and dining experience [C]. INTER-NOISE and NOISE-CON congress and conference proceedings, 2006, 5: 3132-3141.

[7] X CHEN, J KANG. Acoustic comfort in large dining spaces [J]. Applied Acoustics, 2017, 115: 166-172.

[8] CHEN B, KANG J. Acoustic Comfort in Shopping Mall Atrium Spaces—A Case Study in Sheffield Meadowhall [J]. Architectural Science Review, 2004, 47 (2): 107-114.

[9] SHI J H, QIN G, WU J G. Restaurant's lampblack machine and fan noise nuisance case study [J]. Environment and Development, 2013, 25 (11): 153-155.

[10] KANG J. Numerical modelling of the speech intelligibility in dining spaces. Applied Acoustics, 2002, 63 (12): 1315-1333.

[11] WOLFE J. What is acoustic impedance and why is it important? [D]. Pontypridd: University

of New South Wales, 2014: 10.

[12] TO W M, CHUNG A. Noise in restaurants: levels and mathematical model [J]. Noise Health, 2014, 16 (73): 368-373.

[13] YU B, KANG J, MA H. Development of Indicators for the Soundscape in Urban Shopping Streets [J]. Acta Acustica united with Acustica, 2016, 102 (3): 462-473.

[14] LECCESE F, TUONI G, SALVADORI G, et al. An analytical model to evaluate the cocktail party effect in restaurant dining rooms: A case study [J]. Applied Acoustics, 2015, 100: 87-94.

[15] ARIFFIN H F, BIBON M F, ABDULLAH R P S R. Restaurant's atmospheric elements: What the customer wants [J]. Procedia-Social Behavioral Sciences, 2012, 38: 380-387.

[16] LANE H, TRANEL B. The Lombard sign and the role of hearing in speech [J]. Journal of Speech, Language, and Hearing Research, 1971, 14 (4): 677-709.

[17] RONSSE L M, WANG L M. Relationships between unoccupied classroom acoustical conditions and elementary student achievement measured in eastern Nebraska [J]. The Journal of the Acoustical Society of America, 2013, 133 (3): 1480-1495.

[18] HAYNE M J, RUMBLE R H, MEE D J. Prediction of crowd noise [C]. Proceedings of the First Australasian Acoustical Societies Conference, 2006.

[19] HAYNE M J, TAYLOR J C, RUMBLE R H, et al. Prediction of noise from small to medium sized crowds, Acoustics 2011: Breaking New Ground [J]. The Australian Acoustical Society, 2011.

[20] CRISLER B C. The Acoustics and crowd capacity of natural theaters in Palestine [J]. The Biblical Archaeologist, 1976, 39 (4): 128-141.

[21] LONG M. Architectural Acoustics [M]. Elsevier Academic Press, 2006.

[22] COELHO J L B. Community Noise Ordinances, Chapter. 130. 2007: 1525-1532.

[23] LI J N, MENG Q. Study on the soundscape in commercial pedestrian streets [J]. Technical Acoustic, 2015, 34 (6): 326-329.

[24] MENG Q, KANG J. The influence of crowd density on the sound environment of commercial pedestrian street [J]. Science of the Total Environment, 2015, 511 (511C): 249-258.

[25] NIE S S, KANG J. An acoustic model of crowd in large spaces [J]. Journal of Applied Acoustic, 2016, 35 (2): 128-136.

[26] KANG J, MENG Q, JIN H. Effects of individual sound sources on the subjective loudness and acoustic comfort in underground shopping streets [J]. Science of the Total Environment, 2012, 435-436 (7): 80-89.

[27] MILLIMAN R E. The influence of background music on the behavior of restaurant patrons

[J]. Journal of Consumer Research, 1986, 13 (2): 286-289.

[28] WANG L M, VIGEANT M C. Evaluations of output from room acoustic computer modeling and auralization due to different sound source directionalities [J]. Applied Acoustics, 2008, 69 (12): 1281-1293.

[29] WOODS A T, POLIAKOFF E, LLOYD D M, et al. Effect of background noise on food perception [J]. Food Quality and Preference, 2010, 22 (1): 42-47.

[30] FIEGEL A, MEULLENET J F, HARRINGTON R J, et al. Background music genre can modulate flavor pleasantness and overall impression of food stimuli [J]. Appetite, 2014, 76 (5): 144-152.

[31] JIA S Q. Crowd behavior in soundscape [D]. Harbin: Harbin Institute of Technology, 2012.

[32] LINDBORG P M. A taxonomy of sound sources in restaurants [J]. Applied Acoustics, 2016, 110: 297-310.

[33] LINDBORG P M. Psychoacoustic, physical, and perceptual features of restaurants: A field survey in Singapore [J]. Applied Acoustics, 2015, 92: 47-60.

[34] SHENG F C, FANG C Y. Exploring surplus-based menu analysis in Chinese-style fast food restaurants [J]. International Journal of Hospitality Management, 2013, 33 (3): 263-272.

[35] TEMPLE N J, NOWROUZI B. Buffets and obesity [J]. Clinical Nutrition, 2013, 32 (4): 664-665.

[36] WU H C, MOHI Z. Journal of Foodservice Business Research, 2015, 18 (4): 358-388.

[37] HUANG L, ZHU Y, OUYANG Q, et al. A study on the effects of thermal, luminous, and acoustic environments on indoor environmental comfort in offices [J]. Building & Environment, 2012, 49 (1): 304-309.

[38] ZHANG S L, MENG Q. The influence of crowd density on evaluation of soundscape in typical Chinese restaurant [C]. International Conference on Noise and Vibration Control (45th INTERNOISE), 2016. Hamburg, Germany.

[39] BITNER M J. Servicescapes: the impact of physical surroundings on customers and employees [J]. Journal of Marketing, 1992, 56 (2): 57-71.

[40] SHALKOUHI P J. Comments on Reverberation time in an almost-two-dimensional diffuse field [J]. Journal of Sound and Vibration, 2014, 333 (13): 2995-2998.

[41] CAMPBELL C, SVENSSON C, NILSSON E. The same reverberation time in two identical rooms does not necessarily mean the same levels of speech clarity and sound levels when we look at impact of different ceiling and wall absorbers [J]. Energy Procedia, 2014, 78: 1635-1640.

[42] MEISSNER M. Influence of wall absorption on low-frequency dependence of reverberation time in room of irregular shape [J]. Applied Acoustics, 2008, 69 (7): 583-590.

[43] CHOURMOUZIADOU K, KANG J. Acoustic evolution of ancient Greek and Roman theatres [J]. Applied Acoustics, 2008, 69 (6): 514-529.

[44] KANG J. Acoustics of Long space, Theory and Design Guidance [M]. London: Thomas Telford, 2002.

[45] RINDEL J H. Verbal communication and noise in eating establishments [J]. Applied Acoustics, 2010, 71 (12): 1156-1161.

[46] MOON S W, KIM Y J, MYEONG H J, et al. Implementation of smartphone environment remote control and monitoring system for Android operating system-based robot platform [C]. Ubiquitous Robots and Ambient Intelligence (URAI), 8th International Conference on. IEEE, 2011: 211-214.

[47] KARAGEORGHIS C I, JONES L. On the stability and relevance of the exercise heart rate-music-tempo preference relationship [J]. Psychology of Sport & Exercise, 2014, 15 (3): 299-310.

[48] NAVARRO M P N, PIMENTEL R L. Speech interference in food courts of shopping centres [J]. Applied Acoustics, 2007, 68 (3): 364-375.

[49] MARANA A N, VELASTIN S A, COSTA L F, et al. Automatic estimation of crowd density using texture [J]. Safety Science, 1998, 28 (3): 165-175.

[50] ZHANG S L, MENG Q. The influence of crowd density on evaluation of soundscape in typical Chinese restaurant [C]. International Conference on Noise and Vibration Control (45th INTERNOISE), 2016. Hamburg, Germany.

[51] MENG Q, KANG J. Effect of sound-related activities on human behaviors and acoustic comfort in urban open spaces [J]. Science of the Total Environment, 2016, 573: 481-493.

[52] MIGNERON J P, MIGNERON J G. A case study on noise ambience and disturbance in a restaurant [C]. 22nd international congress on sound and vibration, 2015. Florence, Italy.

[53] CHRISTIE L H, BELL-BOOTH R H. Acoustics in the hospitality industry: a subjective and objective analysis, Victoria University of Wellington [C]. New Zealand Centre for Building Performance Research, 2004.

[54] BARRON M. Auditorium acoustics and architectural design [J]. The Journal of the Acoustical Society of America, 1993, 96 (1): 612.

[55] ZAHORIK P. Assessing auditory distance perception using virtual acoustic [J]. The Journal

of the Acoustical Society of America, 2002, 111 (4): 1832-1846.

[56] ZHANG D X, ZHANG M, LIU D P, et al. Soundscape evaluation in Han Chinese Buddhist temples [J]. Applied Acoustics., 2016, 111: 188-197.

[57] YU L, KANG J. Modeling subjective evaluation of soundscape quality in urban open spaces: An artificial neural network approach [J]. The Journal of the Acoustical Society of America, 2009, 126 (3): 1163-1174.

[58] LITWIN M S. How to measure survey reliability and validity [J]. Sage Publications, 1995, 7: 87.

[59] BOUBEZARI M, COELHO J L B. Spatial representation of soundscape [J]. The Journal of the Acoustical Society of America, 2004, 115 (5): 2453-2453.

[60] MENG Q, KANG J, JIN H. Field study on the influence of spatial and environmental characteristics on the evaluation of subjective loudness and acoustic comfort in underground shopping streets [J]. Applied Acoustics, 2013, 74 (8): 1001-1009.

[61] DUBOIS D, GUASTAVINO C, RAIMBAULT M. A cognitive approach to urban soundscapes: Using verbal data to access everyday life auditory categories [J]. Acta Acustica United with Acustica, 2006, 92 (6): 865-874.

[62] KANG J. Urban Sound Environment [M]. London: Taylor and Francis, 2006.

[63] GEORGE D, MALLERY P. IBM SPSS statistics 23 step by step: A simple guide and Reference [M]. Routledge, 2016.

[64] KWOK L, YU B. Spreading social media messages on facebook: An analysis of restaurant business-to-consumer communications [J]. Cornell Hospitality Quarterly, 2013, 54 (1): 84-94.

[65] BECKER G S. A note on restaurant pricing and other examples of social influences on price [J]. Journal of Political Economy, 1991, 99 (5): 1109-1116.

[66] TO W M, CHUNG A W L. Restaurant noise: Levels and temporal characteristics [J]. Noise & Vibration Worldwide, 2015, 46 (8): 11-17.

2.2 Effect of children on the sound environment in fast-food restaurants

2.2.1 Introduction

Restaurants can be divided into five types based on the level of consumption: quick-service restaurants (QSR or fast-food), fast-casual, midscale, moderate (or casual), and fine dining (or upscale)[1]. Of these, fast-food restaurants have gradually become the main choice for eating out, owing to the convenient and fast dining experience that they offer[2]. In selecting fast-food restaurants, customers focus on the restaurant environment[3], particularly the sound environment, as well as factors such as taste[4]. Some previous studies have noted that a restaurant's sound environment significantly affects the customer's will to return and to recommend the restaurant[5]. Another indoor study found that the sound environment (e.g. acoustic comfort) affects a customer's evaluation of their overall satisfaction with the restaurant[6-8]. Another study noted that a restaurant's sound environment affects how people feel about the taste of the food[9]. In terms of sound sources, some studies have noted that the negative evaluation of sound sources in restaurants has a greater impact on potential sources of visitors than positive evaluations[10]. Therefore, it is very important to create a good sound environment in restaurants, particularly fast-food restaurants.

Some previous studies have also demonstrated that the evaluation of the sound environment can be affected by multiple factors (e.g. spaces, human factors and the sound environment itself). Different crowd densities in restaurants have an impact on the sound environment as well as on customers' acoustic comfort[11]. Moreover, while age and gender do not affect the judgement of acoustic comfort, the length of the visit and activities do[12]. The social characteristics of the customer, such as income, education and occupation, also affect acoustic comfort[13]. Different types of restaurant may use different types of background music, which will also have a certain impact on sound comfort[14, 15]. Of these music types, classical concerts encourage people to spend more[16]. The noise of the equipment in the restaurant has an adverse effect on customers' acoustic comfort[17]. People's conversation behaviour at dinner is also affected by the sound environment[18], with, for example, the cocktail party effect, which means people concentrate on interesting conversation and ignore other sounds in restaurants, and this significantly affects customers' acoustic comfort[19]. However, the influence of children's activities on the sound environment and acoustic comfort has not been considered in previous studies.

Although the effects of children's activities on restaurants' sound environment are rarely mentioned, children-directed marketing is being used by more fast-food restaurants[20]. In terms of children's age characteristics, some studies have demonstrated that playing with peers of the same age can effectively eliminate a child's loneliness[21]; thus, many parents take their children to the playground. In terms of gender characteristics, in social problem-solving strategies, girls are more peaceful than boys[22]; the latter expand from small groups into large groups to form peer groups, while girls tend to form individual small groups[23]. There is no obvious difference between boys and girls in terms of communicating with others[24], but girls are more sensitive to the status of people in the group[23]. The gender of the children and their playmates also affects their engagement in activities[25]. Although children have not yet fully developed gender stereotypes, they are more likely to play with children of the same sex[26].

Based on the above, this study aims to ascertain the impact of children's activities in the playground on the SPL and acoustic comfort of fast-food restaurants. First, the influence of the number of children and behaviour combinations while playing on the sound environment were examined. Then, the relationship between children's characteristics and the perception of sound sources were examined. Finally, the effect of children's characteristics on the acoustic comfort of a dining space with a playground was investigated. A typical fast-food restaurant in Harbin city, China, was selected as the case site. The SPL and observation of children's behaviour were measured, while acoustic comfort and sound sources were investigated using questionnaires.

2.2.2 Methods

2.2.2.1 Case Study Site

Numerous previous studies have found that children's playgrounds in fast-food restaurants usually have one of the two typical layouts (Fig. 2.8)[27]: either the playground is located in the dining area or it is located outside of the dining area. In the first arrangement, there is usually no wall between the children's playground and the dining area so that sound transmission cannot be effectively prevented[28]. Under this arrangement, the children's playground has a greater effect on the sound environment of the dining area. To investigate the influence of children's playgrounds on the sound environment, this paper selected a case study site featuring the first arrangement.

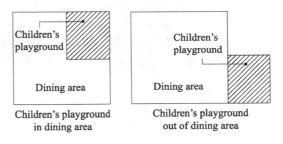

Fig. 2.8 **Two types of layout mode**

A Kentucky Fried Chicken (KFC) outlet in Daoli District, Harbin, China, was selected as the study site. Within this KFC outlet, furniture divides the second floor into two main dining areas, one of which contains a children's playground (Fig. 2.9). There are 13 tables in this dining area, which can accommodate 33-40 persons. The children's playground is located in one corner of the dining area, with an area of approximately 14m², accounting for 20% of the dining area. A 1.2-meter-high fence divides the children's playground from the dining area. The decoration materials of the dining area and children's playground can be found in Table 2.3.

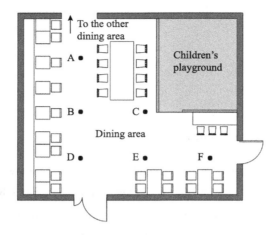

Fig. 2.9 Plan of case site

Table 2.3 Interior materials of ceilings, walls and floors in the case study site

Study site		Children's playground	Dining area
Photos			
Interior materials and sound absorption coefficient	Ceilings	Gypsum $\alpha = 0.3$	Gypsum $\alpha = 0.3$
	Walls	Soft pack $\alpha = 0.5$	Plastic $\alpha = 0.08$
	Floors	Cystosepiment $\alpha = 0.23$	bricks $\alpha = 0.01$
	Tables		Wood $\alpha = 0.03$

2.2.2.2 Observation of Children's Behaviour

After gaining permission from their parents, an HD video camera was used to record the behaviours of children playing in the playground. Observations of the children's characteristics in the case site were carried out every weekend and lasted one year (from Nov. 2017 to Oct. 2018) because of the following reasons. It was found that there was no significant difference in the number and behaviour of users between weekdays and weekends at KFC, as well as SPL and the types of background music. However, as children usually go to the case site at the weekend, it is easier to investigate the different number of children at the weekend. As our previous investigation had shown that this is the time of highest usage of the playground, the observation period lasted from 10:00 am to 6:00 pm. Children aged between three and six years were selected for observation, as previous studies have demonstrated that this age group can operate independently in the playgrounds[29]. Furthermore, for safety reasons, the children's playground does not recommend usage by very young children, while children over six years gradually decrease their interest in this kind of play and instead prefer games that challenge their intelligence[30].

As children require some time to adapt to the playground environment[31], the observation began at least one minute after the child had entered or exited the playground. Some previous studies have demonstrated that children's concentration increases with age[32]. In the observation, children in the playground could usually maintain stable activity for eight to ten minutes. Therefore, the observation time for each playing types was set at five minutes. During the observation, children sometimes suddenly ran out of the playground. In such cases, observation continued so long as the child was not separated from the activities in the playground. If the child disengaged from the activities, however, the observation would cease.

According to the observation of children's performance in the activities, the combination of their activities can be divided into three basic types: centralised, partially centralised and dispersal. The centralised combination means that all children in the field form a group to play together. The dispersal combination means that no playgroup is formed; instead, every child takes care of himself and plays alone. It should be explained that the partially centralised combination means that some of the children in the playground form a playgroup, while other children play alone.

2.2.2.3 Measurement of acoustic environment

SPL was measured at the same time as the observation of the children's behaviour. The data lasting for 3-5min and the interval between the two data was at least 1min. The sound-level meter was set 1m away from the wall and main reflector and 1.2-1.5m from the ground[33]. The data were measured every second using fast-style and A-weight[34]. At the survey site, to avoid measurement errors, six measurement points (A-F) were first selected in the initial SPL measurement, and every point was at least 2m apart. The results of each point were not significantly different from the other points[35]. Therefore, point C was selected in the formal investigation, to avoid the influence of the users, as fewer users sit nearby the children's playground. In order to avoid the influence of people sitting close to the measurement point, the measurement was carried out only when there was no user nearby.

In the analysis, we took the data average as the average SPL and the instantaneous max value as the max SPL and the min value as the Min SPL, because the children's sudden shouts are at a high frequency and a high decibel level, which may influence both the instantaneous SPL and acoustic comfort. Simultaneously, the duration of each shout or loud communication was relatively short. Taking the instantaneous max value as the Max SPL can reflect the change in the sound environment more truly and accurately.

As reverberation time may influence the sound feeling in an indoor space[36, 37], we also took this factor into account. Based on the investigation, the number of people in the dining area was generally between 6 and 35. From the calculation using the Eyring formula, the T30 (500Hz) was from 0.52s (6 users) to 0.40s (35 users). The difference in the reverberation time (RT) was only 0.12s, and therefore, the effect of RT based on different crowd density was not considered in this paper[18].

2.2.2.4 Questionnaire survey

In this section, we introduce the length of survey time, selection and criteria for volunteer, acoustic comfort survey and investigation of sound sources.

1. Length of survey time

After initial observations at the case site, it was found that people in fast-food restaurants engage in various activities such as having a brief rest, waiting for meals, dining, and resting after dinner[18], although the main activity is eating. The length of time spent in fast-food restaurants also varies[38]. After more than 850 minutes of observation, the main behaviour affecting stay time is rest, with people in fast-food restaurants usually spending between 10-25min eating, most of which is concentrated in approximately 15min. Therefore, this study selected 15min as the period of time within which to investigate the sound source and acoustic comfort.

2. Selection and criteria for volunteers

The users at the case site can be generally divided into two groups. The first group includes parents who have a relationship with the children, and the second group includes people who are not related to the children. From the first group, the relationships between the parents and children may influence their evaluation of a sound environment. Meanwhile, teenagers and young adults are the typical service object of fast-food restaurants, and they usually have higher expectations for the restaurant environment. That being the case, 25 university volunteers (aged 20-24 years) were selected, with relatively equal numbers of females (12) and males (13). Since some studies have noted that sound perceptions may vary by gender[39], the genders of volunteers were balanced to avoid the influence of gender. Participants were asked to investigate the sound source and acoustic comfort in the restaurant by filling out questionnaires. They were just told the aims of this project were to record sound sources and acoustic comfort, to ensure that they did not pay too much attention to the children's voice. After arriving at the case site, they were given 5min to adapt to the new environment to allow them to settle and judge the environment relatively accurately. To avoid habituation caused by a long stay, each participant recorded observations for no more than an hour. To ensure that the fast-food restaurant experience was authentic and that participants' feelings replicated real-life while conducting the questionnaire, volunteers were free to engage in all activities usually conducted in the KFC outlet, such as dining, playing on their mobile phones, and chatting with friends. Then, the number of each sound source in the questionnaire and the percentage of the sound source to the total sound source was given in each period. In this way, the impact of the sound source on the volunteers was expressed.

3. Acoustic comfort survey

A five-level scale was used to survey acoustic comfort[40]: 1, very uncomfortable; 2, uncomfortable; 3, neither uncomfortable nor comfortable; 4, comfortable; and 5, very comfortable[41]. During the survey period, volunteers were asked to record the time and acoustic comfort, as well as the factors influencing their evaluation of comfort. According to the changing characteristics of the children's behaviour in the playground, it was determined that volunteers would generally be required to record randomly at 4-6 time points every 15 min. The selection of such intervals ensured that there would be sufficient records in each period, and avoided the effect of too many subjective surveys on the results of the records. In the analysis, the average value of each data point was used to indicate the acoustic comfort of the period.

4. Investigation of sound sources

In the survey, volunteers were asked to record the sound sources they heard at each time point. Repeated sound sources need to be recorded repeatedly, while at the same time recording the sound source that had the greatest impact on them. Based on an investigation of more than 200 persons (in 15 days, accumulated), which were people dining in the fast-food restaurant, 13 typical sound sources in fast-food restaurants with a children's playground were summarised[27]. In this study, all other sources, such as customers' chewing sounds, accounted for less than 1% and were excluded from the analysis. In the questionnaire, the sound sources were divided into sound source from the restaurant and sound source associated with the playground (Table 2.4). In this study, 20s was set as one period in the survey, because during this period, volunteers have enough time to record what they hear. Some sources are continuous, and these were considered as occurring once within 20s. There is a distinction between the sound of communication between customers and the sound of communication between parents because customers and parents who accompany children usually have different behavioural patterns and have different effects on acoustic comfort. These criteria were used as data to study the effect of sound sources on acoustic comfort.

Table 2.4 The main sound sources in the case site

Code name	Types of sound	Code name	Types of sound
a	Background music	h	Footsteps
b	Sound of equipment	i	Sound of children beating an instrument
c	Sound of cleaning	j	Sound of children screaming and crying
d	Sound of table and chair friction	k	Sound of communication among children
e	Sound of communication between staff and customers	l	Sound of communication among children and parents
f	Sound of communication between customers	m	Sound of communication among parents
g	Sound of customers playing games		

2.2.2.5 Statistics and analysis

SPSS 24.0[42] was used to establish a database with all the subjective and objective measurements. In this study, linear and nonlinear regression analyses were used to calculate the relationship between the number of children and SPL, as well as among SPL, sound sources and acoustic comfort.

2.2.3. Results

2.2.3.1 Effect of children on SPL

In this section, the effect of the number and behavioural combinations of children on the sound environment is explored.

1. Number of children

The average SPL of the background noise in the dining area was 66.3dB(A) (standard deviation 2.1) when there were no children in the playground. When there were no children or only one child in the field, children had little effect on SPL, with T-test, $p > 0.1$. Due to the size of the playground, it was rare for more than eight children to play there at the same time. Therefore, this study investigates numbers of between two and eight children.

A multiple regression analysis was used to analyse the data. The effect of the number of children on SPL was found to be the highest fitting degree with the cubic curve, $R^2 = 0.992$, $p < 0.001$. As shown in Fig. 2.10, the change in SPL first increases and then tends to stabilise. When there are fewer than four children, SPL demonstrates a trend of rapid increase. The average SPL is 70.6dB(A) when there are two children and increases to 76.2dB(A) when there are four children. On average, for every additional child, SPL increases by 2.8dB(A). When there are more than four children in the playground, the change of SPL tends to stabilise, with an average SPL of 76.8dB(A) and fluctuating by 0.5dB(A).

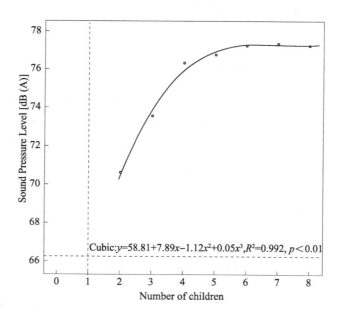

Fig. 2.10 **Effect of the number of children on the average SPL**

After analysing the fluctuation range of SPL for each number of children, the Min SPL was found to increase gradually with the increase in the number of children. With an increase of one child, SPL increases by 1.8dB(A) on average. The Max SPL does not keep increasing. As seen in Fig. 2.11, the Max SPL usually exceeds 100dB(A) with fewer than five children. In particular, when there are only three children in the playground, the Max SPL reaches nearly 120dB(A). When the number exceeds five, the Max SPL decreases significantly. The average Max SPL is 94.7dB(A).

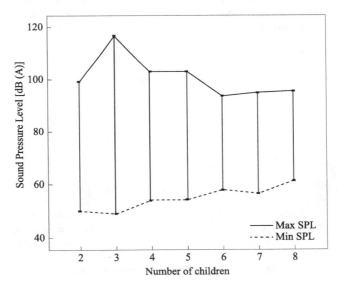

Fig. 2.11 The effect of the number of children on the Max and Min SPL

2. Behavioural combinations of children

When there are two children, there are only two possible combination types in the playground: centralised and dispersal. When there are seven or more children, almost no centralised combinations were observed. Therefore, the relationship among the number, the combinations and SPL was analysed for cases of three to six children.

When there are three children, there is a 67% possibility of forming a centralised combination, with only 16.5% possibility of forming a dispersal combination (Fig. 2.12). As the number increases, the possibility of forming a dispersal combination gradually increases. When there are six children, the

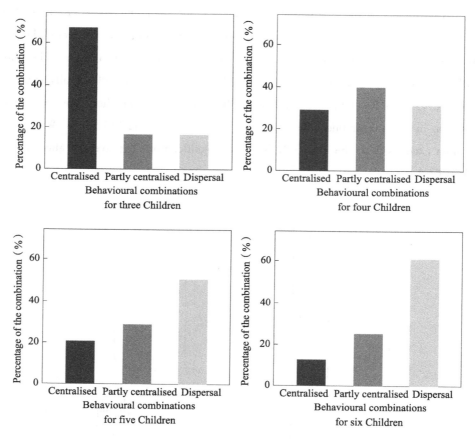

Fig. 2.12　Percentage of each combination

probability of occurrence of the centralised combination is reduced to 13%, and the probability of dispersal combination increases to 61%. The probability of a partially centralised combination decreases with the increase in the number of children. The highest probability of occurrence is with four children, at up to 40%. This combination is widely observed when there are three or more children in the playground.

When the combination is centralised, children need to communicate through language to form and maintain playgroups. Simultaneously, with an increase in the number of children, the difficulty of maintaining a playgroup increases. Children of this age are usually at a stage where language and organisational skills are just beginning to develop[43, 44]. Thus, it is difficult for them to calmly face the communication challenges posed by a greater number of children, which would require them to communicate in a louder voice or with more intensive language to maintain a stable presence in the group. When there are fewer than four children, the centralised combination prevails. In combination with Fig. 2. 10, this demonstrates that the average SPL increases rapidly as the number of children increases. The trend of the Max SPL changes in Fig. 2. 11 is also related to the combinations. When the number exceeds four, the dispersal combination gradually prevails. In this combination, children no longer require frequent language communication, which stops the average SPL from increasing. When no children suddenly screamed or communicated in the field, most children were able to remain relatively quiet while playing.

It is speculated that when the number of children exceeds eight, SPL will remain relatively stable. This is because, according to the study of children's behavioural combinations, it can be inferred that more children will lead to a larger percentage of dispersal combinations and communication between children will remain at a low frequency.

2. 2. 3. 2 Effect of children on sound sources

As the sound level effect can be different, in this section, the effect of children is explored. As seen in Fig. 2. 13, the percentage of sound sources associated with the playground changed with the number of children. As a whole, however, the percentage of each sound source fluctuated within a certain range, and there was no sudden increase or decrease. Overall, the two sources with the largest percentage are j-sound (sound of children screaming and crying) and k-sound (sound of communication among children), while the sources with the lowest percentage are l-sound (sound of communication among children and parents) and m-sound (sound of communication among parents).

In terms of the sound of children beating instruments, the percentage of i-sound (sound of children beating instrument) remained stable with an average of 17%. The slide in the playground includes parts that children can tap and rotate to make a sound. In the observation, regardless of the number of children, they were attracted to this setting. With the increase in the number of children, there were some

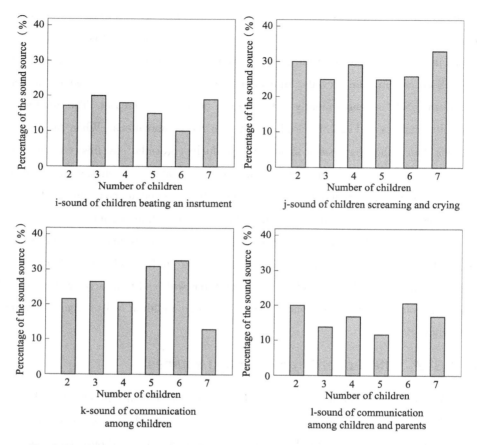

Fig. 2.13 Percentage of varied sound sources with different number of children

fluctuations in the percentage of i-sound, which may be related to the increase in other sound sources.

In terms of the sound of children screaming and crying, the average percentage of the j-sound is 28% (from 25% to 33%), and this fluctuated with the increase in the number of children. Children's screams and cries were often associated with contests over amusement equipment, their heads colliding with instruments while playing, and child-to-child bickering, and so on. It was found that although there was a low probability of these conditions occurring, the proportion of this sound source continued to be high. This phenomenon may be related to this sound source, which can easily attract people's attention and has a great influence on acoustic comfort. A detailed analysis will be conducted in section 1.

In terms of the sound of communication between children, the percentage of k-sound first increased and then decreased with the increase in the number of children. When there were five or six children, the percentage was the highest, at 31% or 33%. The analysis in section 2 demonstrates that when there are fewer children, they tend towards the centralised combination, which they need to maintain through language communication. However, as the number of children increases, more language communication is required. When there are a large number of children, the combination is mainly of the dispersal type, which does not require very much language communication to maintain. As a result, this sound source presents the changes illustrated in Fig. 2.12.

In terms of the sound of communication between children and parents, the percentage of l-sound first decreased from 20% (for two children) to 12% (for five children) and then increased to 17% (for seven children). In the observations, parents had a higher probability of communicating with children outside the playground when the number of children was small or large. However, the purpose of communication in these two situations differs. When there were few children, the communication was more likely to be a pretend game to enliven the atmosphere in the playground. When the number was large, the communication was more likely to be to maintain order in the playground.

2.2.3.3 Effect of children's related sound environment on acoustic comfort

In this section, the effect of sound level and sound sources on acoustic comfort is explored, to determine how to control the number of children to improve the acoustic comfort in a fast-food restaurant.

1. Effect of SPL on acoustic comfort

Some previous studies have pointed out that SPL in a fast-food restaurant is generally no more than 80dB(A)[45]. However, in the present study, sounds of more than 80dB(A) were usually related to children, which may influence the evaluation of the acoustic comfort of diners. While with an increase to the Max SPL, the acoustic comfort was significantly reduced. However, due to the distribution of the Max SPL, it is difficult to judge the correlation between acoustic comfort and the Max SPL. Therefore, in this section, the relationship between acoustic comfort and the duration of the Max SPL exceeding 80dB(A) is analysed within 15min. The following paragraph uses the OVER 80 to represent the Max SPL that exceeds 80dB(A). The duration of OVER 80 is taken as the basis of analysis. For example, if the OVER 80 is measured for 47s in 15min, its duration is in the 30-60s group.

In Fig. 2.14, the regression analysis shows that there is a significant linear correlation between acoustic comfort and the duration of OVER 80, with $R^2 = 0.880$, $p < 0.001$. When the duration of OVER 80 is 0, the acoustic comfort is 3.4. With the increase in duration, acoustic comfort decreases. When it reaches 1.5min, the acoustic comfort is 2.5. The average acoustic comfort decreased by 0.3 every 30s. When the duration is between 1.5min and 4min, the acoustic comfort shows a wave. The average acoustic comfort is 2.6, with the highest acoustic comfort at 2.9 and the lowest at 2.2. Subsequently, the acoustic comfort demonstrated a decreasing trend with the increase in duration. Despite some fluctuations, it was maintained at approximately 1.6. Once the duration exceeded 7min, the acoustic comfort decreased to 1.2, which is close to a very uncomfortable degree.

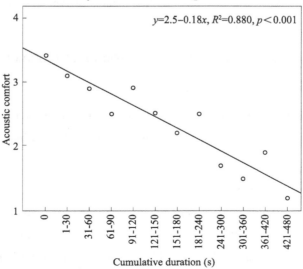

Fig. 2.14 Effect of duration time of SPL exceeding 80dB(A) on acoustic comfort

When the duration is less than 1.5min, much of OVER 80 occurs in the form of a sudden rise [Fig. 2.15 (a)]. A sudden rise in SPL can easily elicit negative emotions, such as irritability and shock. As a result, the acoustic comfort decreased significantly at this stage. When the cumulative duration is 1.5-4min, OVER 80 will appear more continuously [Fig. 2.15 (b)]. Although the duration increases, the sudden increase in SPL decreased. Therefore, at this stage, although the acoustic comfort fluctuates, the overall trend is stable. On the other hand, when the duration exceeds 4min, OVER 80 continues to appear [Fig. 2.15 (c)], and the influence of SPL on acoustic comfort becomes more obvious.

Combined with Figs. 2.10 and Figs. 2.11, the findings demonstrate that when there are fewer children, the Max SPL is higher, but the average SPL is lower, and the possible duration of OVER 80 is shorter. When there are more children, the opposite is true. In this case, the Max SPL is relatively low, but the average SPL is higher, indicating that the possible duration of OVER 80 is longer. Fig. 2.16 illustrates the effect of the number of children on average acoustic comfort of participants. As shown in Fig. 2.16 when the number of children was four or six, the average acoustic comfort of participants was maintained at about 2.6. However, when the number of children increased to seven or eight, the acoustic comfort decreased to between 1.7 and 1.8.

2. Effect of sound sources on acoustic comfort

An interesting phenomenon is found by the statistical analysis of the most influential sound sources. Thirteen common sound sources were listed in the questionnaire (Table 2.4), but only nine of the sound sources in the item "Sound sources that have the greatest impact on you". Four of the nine sound sources came from the restaurant, accounting for 23%; and five sound sources were related to the playground, accounting for 77%. This result demonstrates intuitively that in this space, the influence of the sound source related to the playground is not negligible. There are also obvious differences in the percentages of the nine sound sources. As seen in Fig. 2.17, the largest percentage is the sound source j-sound, accounting for 38%. The second-largest is the sound source k-sound, at 24%. The percentage of sound sources a-sound (background music) and f-sound (sound of communication betwcen customers) is significantly lower than the first two items, at 10% each. The other five sound sources account for an average of 4%.

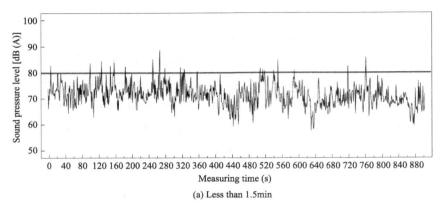
(a) Less than 1.5min

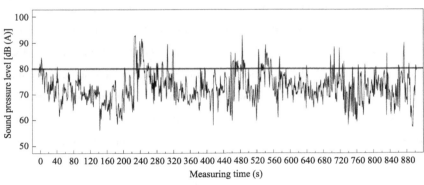
(b) Between 1.5min and 4min

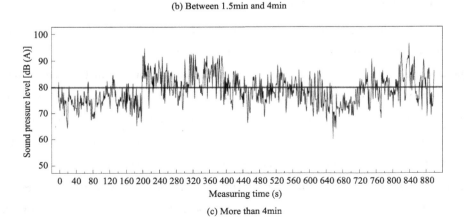
(c) More than 4min

Fig. 2.15 Duration of OVER 80

Chapter 2　Sound Environment and Acoustic Perception in Dining Spaces

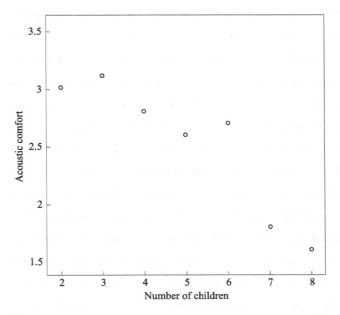

Fig. 2.16　Effect of the number of children on acoustic comfort

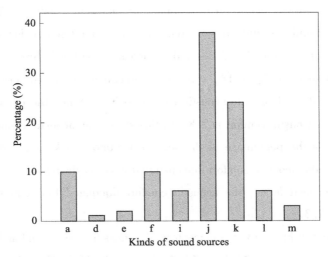

Fig. 2.17　Percentage of various sound sources

Note: a—background music; d—sound of table and chair friction; e—sound of communication between staff and customers; f—sound of communication between customers; i—sound of children beating an instrument; j—sound of children screaming and crying; k—sound of communication among children; l—sound of communication among children and parents; m—sound of communication among parents.

Although a-sound and f-sound have a relatively large influence on people, the correlation test demonstrated that they have no significant correlation with acoustic comfort, at $p > 0.01$. Simultaneously, there is no significant correlation between acoustic comfort and other sound sources from the restaurant, at $p > 0.01$. Thus, although some of the sound sources from the restaurant may have some influence on acoustic comfort, they will not lead to a significant decrease or increase in acoustic comfort in the fast-food restaurant with a playground set-up. With the change in the number of children, there is no significant change in sound sources associated with the playground: i-sound, l-sound, and m-sound. Meanwhile, there is no significant correlation between these sound sources and acoustic comfort ($p > 0.01$).

The analysis demonstrated that the single source j-sound or k-sound has little effect on acoustic comfort, but the combination of the two sources has an obvious effect on acoustic comfort. This may be because each sound source appears more dispersed when the effects on sound comfort of these two sound sources are calculated separately, and the two sound sources appear more continuous after the two sources are combined, which has a greater impact on acoustic comfort. Therefore, the sources j-sound and k-sound are combined into one sound source for analysis. The effect of the sources j-sound and k-sound on acoustic comfort is linear, $R^2 = 0.827$, $p < 0.05$. As seen in Fig. 2.18, when the percentage of the sources j-sound and k-sound is less than 20%, the acoustic comfort is 3.3. When the percentage is less than 30%, the acoustic comfort is 2.8. At this stage, the acoustic comfort decreased significantly. As the percentage of the sources j-sound and k-sound increased from 30% to 60%, the acoustic comfort demonstrated a slowly decreasing trend. Once the percentage exceeded 60%, the acoustic comfort fluctuated but remained stable at approximately 2.3.

As observed in Fig. 2.13, the percentage of sources for the j-sound and k-sound first increases and then decreases slightly with the increase in the number of children, with a percentage of 50%-59% for five to seven children. In combination with Fig. 2.18, it can be inferred that with the increase in the number of children, the acoustic comfort decreases gradually. This result was also consistent with the results in Fig. 2.16.

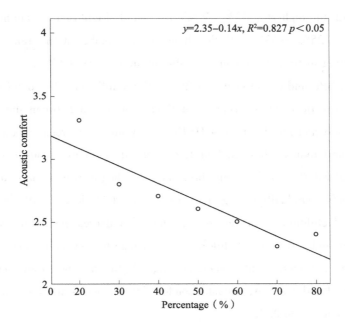

Fig. 2.18 Effect of the sources j-sound and k-sound on acoustic comfort

2.2.4 Conclusions

Based on measurements, observations and questionnaires, this study has examined the influence of a children's playground on the sound environment in a fast-food restaurant. The conclusions are as follows.

First, the change in the average SPL is related to the number of children, but not linearly. As the number of children increases, the average SPL increases rapidly and then stabilises. A total of four children is the boundary point between two changing trends. When there are fewer than four children, SPL increases the fastest at 2.8dB(A) per child; the Min SPL increases as the number of children increases, with the average of 1.8dB(A) per child; and the Max SPL also has four children as the boundary point. When there are fewer than four children, the Max SPL retains a higher level of more than 100dB(A) and then decreases significantly.

Second, some typical sounds, such as the sound of communication among children, first increased and then decreased with the increase in the number of children, with the highest at 33%. While the sound of children screaming and crying fluctuated from 25% to 33% with the change in the number of children, the sound of children beating instruments remained stable at an average of 17%.

Finally, SPL and sound sources can effectively influence the sound environment in terms of acoustic comfort. There is a linear correlation between the variation of acoustic comfort and the duration of OVER 80. With the increase in the Max SPL, acoustic comfort reduced from 3.4 to 1.2. The effect on the acoustic comfort of a duration of OVER 80 is divided into three stages. It begins with a sharp decline, then fluctuates, before gradually falling to a very uncomfortable level. With the increase in the number of children, acoustic comfort showed a decreasing trend, with 2.1 for three children and 1.6 for eight children. The change to acoustic comfort is correlated with children screaming and communicating. With the increased percentage of sources j-sound and k-sound, acoustic comfort first decreases and then stabilises with the change from 3.3 to 2.3.

The limitations of this study are, first, that it only examined a restaurant with a fixed children's playground, and of the fast-food restaurant type. The maximum number of children in the playground was ten, and the number studied in this paper ranged between two and eight children aged three to six. Thus, this paper does not cover cases of more than eight children. In terms of gender combinations, this study examined gender combinations of three and four. A different number of children and different gender combinations, and so on, may have different effects on the sound environment, and this will be examined in future studies.

References

[1] CANZIANI B F, ALMANZA B, FRASHJR R E, et al. Classifying restaurants to improve usability of restaurant research [J]. International Journal of Contemporary Hospitality Management, 2016, 28 (7).

[2] DONG X, HU B. Regional difference in food consumption away from home of urban residents: a panel data analysis [J]. Agriculture and Agricultural Science Procedia, 2010, 1 (1): 271-277.

[3] ARIFFIN H F, BIBON M F, ABDULLAH R P S R. Restaurant's atmospheric elements:

what the customer wants [J]. Procedia-Social and Behavioral Sciences, 2012, 38: 380-387.

[4] STEWART H, BLISARD N, BHUYAN S, et al. The demand for food away from home: full-service or fast-food? [J] Agricultural Economics Reports, 2004.

[5] HEUNG V C S, GU T M. Influence of restaurant atmospherics on patron satisfaction and behavioral intentions [J]. International Journal of Hospitality Management, 2012, 31 (4): 1167-1177.

[6] CHEN X, KANG J. Acoustic comfort in large dining spaces [J]. Applied Acoustics, 2017, 115: 166-172.

[7] WANSINK B, VAN I K. Fast food restaurant lighting and music can reduce calorie intake and increase satisfaction [J]. Psychological Reports, 2012, 111 (1): 228.

[8] SPENCE C. Noise and its impact on the perception of food and drink [J]. Flavour, 2014, 3 (1): 9.

[9] CARVALHO F R, VAN EE R, RYCHTARIKOVA M, et al. Using sound-taste correspondences to enhance the subjective value of tasting experiences [J]. Frontiers in Psychology, 2015, 6: 1309.

[10] XI W, LIANG R, TANGA E K. More than words: do emotional content and linguistic style matching matter on restaurant review helpfulness? [J]. International Journal of Hospitality Management, 2018, 77: 438-447.

[11] ZHANG S, MENG Q, KANG J. The influence of crowd density on evaluation of soundscape in typical Chinese restaurant, Inter-noise and Noise-Con Congress and Conference Proceedings, 2016.

[12] CHEN B, KANG J. Acoustic comfort in shopping mall atrium spaces: a case study in Sheffield Meadowhall [J]. Architectural Science Review, 2004, 47: 107-114.

[13] MENG Q, KANG J. Influence of social and behavioural characteristics of users on their evaluation of subjective loudness and acoustic comfort in shopping malls [J]. PLOS ONE, 2013, 8 (1): e54497.

[14] HANNAH L. Sound and the Restaurant Environment [M]. McGraw-Hill, 2004.

[15] WANSINK B, ITTERSUM K V. Fast food restaurant lighting and music can reduce calorie intake and increase satisfaction [J]. Psychological Reports, 2012, 111 (1): 228.

[16] NORTHA C, SHILCOCK A, HARGREAVES D J. The effect of musical style on restaurant customers's pending [J]. Environment and Behavior, 2003, 35 (5): 712-718.

[17] SHI J, QIN G, WU J. Restaurant's Lampblack machine and fan noise nuisance case study [J]. Northern Environment, 2013.

[18] MENG Q, ZHANG S, KANG J. Effects of typical dining styles on conversation behaviours and acoustic perception in restaurants in China [J]. Building and Environment, 2017, 121: 148-157.

[19] LECCESE F, TUONI G, SALVADORI G, et al. An analytical model to evaluate the cocktail party effect in restaurant dining rooms: a case study [J]. Applied Acoustics, 2015, 100: 87-94.

[20] OHRI-VACHASPATI P, ISGOR Z, RIMKUS L, et al. Child-Directed Marketing Inside and on the Exterior of Fast Food Restaurants [J]. American Journal of Preventive Medicine, 2015, 48 (1): 22-30.

[21] ZHOU Z, SUN X, XIANG Y, et al. Peer interaction and loneliness in middle childhood: a cross-lagged analysis [J]. Psychological Science, 2007, 30 (6): 1309-1313.

[22] WALKER S, IRVING K, BERTHELSEN D. Gender influences on preschool children's social problem-solving strategies [J]. The Journal of Genetic Psychology, 2002, 163 (2): 197-209.

[23] BENENSON J, APOSTOLERIS N, PARNASS J. The organization of children's same-sex peer relationships [J]. New Directions for Child and Adolescent Development, 2010, 1998, 81: 5-23.

[24] LEAPER C. Influence and involvement in children's discourse: age, gender, and partner effects [J]. Child Development, 2010, 62 (4): 797-811.

[25] FABES R A, MARTIN C L, HANISH L D. Young children's play qualities in same-, other-, and mixed-sex peer groups [J]. Child Development, 2010, 74 (3): 921-932.

[26] HAMZA A. Children's play behavior in an outdoor play setting: case study in Hadana Pusat Islam, U. S. M. Chinese Research Institute of Construction Management, 2010: 537-547.

[27] LIU S, MENG Q, KANG J. Effects of children characteristics on sound environment in fast-food restaurants in China [J/OL]. Euronoise 2018. https: //www.researchgate.net/publication/325550433_Effects_of_children_characteristics_on_sound_environment_in_fast_food_restaurants_in_China.

[28] MENG Q, KANG J. The influence of crowd density on the sound environment of commercial pedestrian streets [J]. Science of The Total Environment, 2015, 511: 249-258.

[29] MASHBURN A J, PIANTA R C, HAMRE B K, et al. Measures of classroom quality in prekindergarten and children's development of academic, language, and social skills [J]. Child Development, 2008, 79 (3): 732-749.

[30] DWYER G M, BAUR L A, HARDY L L. The challenge of understanding and assessing physical activity in preschool-age children: Thinking beyond the framework of intensity, duration and frequency of activity [J]. Journal of Science and Medicine in Sport, 2009,

12 (5): 534-536.

[31] WASIK B A, JACOBI-VESSELS J L. Word play: scaffolding language development through child-directed play [J]. Early Childhood Education Journal, 2017, 45 (6): 769-776.

[32] BELL E R, GREENFIELD D B, BULOTSKY-SHEARER R J. Classroom age composition and rates of change in school readiness for children enrolled in Head Start [J]. Early Childhood Research Quarterly, 2013, 28 (1): 1-10.

[33] YANG W, KANG J. Acoustic comfort evaluation in urban open public spaces [J]. Applied Acoustics, 2005, 66 (2): 211-229.

[34] CHRISTIE L H, BELL-BOOTH R H. Acoustics in the hospitality industry: a subjective and objective analysis, Victoria University of Wellington. New Zealand Centre for Building Performance Research, 2004.

[35] JAFFE B F, LAWRENCE M. The dynamic range of hearing [J]. Surgical Forum, 1966, 17: 464.

[36] ZANNIN P, TROMBETTA H, MARCON C C R. Objective and subjective evaluation of the acoustic comfort in classrooms [J]. Applied Ergonomics, 2007, 38 (5): 675-680.

[37] MEISSNER M. Influence of wall absorption on low-frequency dependence of reverberation time in room of irregular shape [J]. Applied Acoustics, 2008, 69 (7): 583-590.

[38] NOONE B M, KIMES S E, MATTILA A S, et al. The effect of meal pace on customer satisfaction [J]. Cornell Hospitality Quarterly, 2007, 48: 231-244.

[39] DEPISAPIA N, BORNSTEIN M H, RIGO P, et al. Sex differences in directional brain responses to infant hunger cries [J]. NeuroReport, 2013, 24 (3): 142-146.

[40] YU L, KANG J. Modeling subjective evaluation of soundscape quality in urban open spaces: an artificial neural network approach [J]. Journal of the Acoustical Society of America, 2009, 126 (3): 1163-1174.

[41] KANG J, MENG Q, JIN H. Effects of individual sound sources on the subjective loudness and acoustic comfort in underground shopping streets [J]. Science of the Total Environment, 2012, 435-436: 80-89.

[42] GEORGE D, MALLERY P. IBM SPSS Statistics 23 Step by Step: A Simple Guide and Reference [M]. Routledge, 2016.

[43] PIAGET J, MAYS W. The principles of genetic epistemology [J]. Philosophical Quarterly, 1972, 24 (94): 87.

[44] JIANG M. An initial research for nonverbal communication development of children aging from 0 to 6 [J]. Chinese Journal of Special Education, 2004, 5 (47): 79-84.

[45] TO W M, CHUNG A. Noise in restaurants: levels and mathematical model [J]. Noise and Health, 2014, 16 (73): 368.

Chapter 3

Sound Environment and
Acoustic Perception
in Railway Stations

3.1 Acoustic comfort in large railway stations

3.1.1 Introduction

Railway stations have traditionally been associated with waiting and transit spaces. In the past, this association was because the stations hosted a relatively limited number of functions[1]. Currently, however, large railway stations worldwide are being built to accommodate increasingly complicated functions and crowds, which has introduced more stringent requirements for the sound environment. Acoustic comfort, which is the most important index for evaluating soundscape[2], has also been widely studied in public spaces, including offices[3], large dining rooms[4], public libraries[5], commercial spaces[6], quiet and restorative areas[7,8]. In these studies, transportation noise sources are generally mentioned as the primary or secondary noise sources. Researchers have found that different types of vehicles have a specific impacts on the surrounding environment, for instance, noise from road traffic[9], trains[10], aircraft[11] and vessels[12]. These noises have been demonstrated to make diffuse people and disturb them in[13] residential areas[14], commercial areas[15], school areas[16], quiet natural areas[17] and port areas[18]. Several adverse effects have been associated with exposure to traffic noise[19]. Beyond its effects on the auditory system, noise causes annoyance[20], disturbs sleep[21] and impairs cognitive performance[22]. Furthermore, epidemiologic studies have demonstrated that environmental noise is associated with increased arterial hypertension, myocardial infarction, and stroke[23]. Moreover, aircraft and road traffic noise exposure have been associated with psychological symptoms[24]. In children, chronic aircraft noise exposure impairs reading comprehension and long-term memory and may be associated with increased blood pressure[25]. Generally, it has been found that continuous exposure to traffic noise causes people to suffer from various types of discomfort and appreciably reduces measures of well-being. However, little attention has been paid to people's levels of acoustic comfort inside transit spaces. People's comfort and psychophysical well-being are important in transit spaces and should be significant considerations during the designs of such spaces; however, further details concerning the phenomena and theory are still required.

In China's railway stations, the entrance hall, ticket office, integrated waiting hall and auxiliary space are all concentrated within a single large space[26]. As the number of functions increases, the types of sound sources also increase. The resulting complex acoustic environment leads to various adverse effects on user comfort and causes a series of acoustic problems, such as high environmental noise and poor language articulation[27]. Conversations between people are a primary behavioral factor influencing the sound environment and acoustic perceptions in railway stations. Studies have found that noise emissions from activities involving crowds of people can also affect the sound environments of public spaces[28]. Nie and Kang[29] proposed a crowd acoustic model and found a relationship between the crowd, sound pressure level, total population and number of people conversing. Wu and Kang[30] using the results of interviews and questionnaires, showed that conversational speech intelligibility is poor and that crowd noise is considered a main factor that affects broadcast clarity. Traffic sounds are another main sound source in railway stations. Bandyopadhyay et al.[31] measured the sound pressure level (SPL) on platforms and found that SPL affects users' health. Broadcasts are an important aspect of the sound in railway stations. Liu et al.[32] used acoustic measurements and simulations to study the reverberation time (RT) and the speech transmission index of public broadcasting systems. Excessive noise exposure caused by the enormous ventilation systems in large spaces also has serious impacts. Tao et al.[33] evaluated the noise annoyance levels in a metro depot and the noise influence of its ventilation system on adjacent residential buildings. However, simply reducing the overall 'sound level' does not always result in the desired quality of life improvements. Many sound sources have been studied to evaluate their influences on the sound environment and acoustic comfort, but due to their increasing functions, the acoustic comfort levels in large railway stations have not been studied systematically.

Chapter 3 Sound Environment and Acoustic Perception in Railway Stations

Therefore, the goal of this chapter is to study the effects of various sound sources on the sound environment and acoustic comfort in extra-large spaces using a soundscape approach. A typical large railway station in China was chosen as a case study. The overall comfort level and sound environment in the station's different functional zones were studied using both objective measurements and a questionnaire survey. First, the sonic composition and appropriateness were analyzed. Then, the effects of different types of sound sources in different zones on loudness, intelligibility, sound level, preference degree and acoustic comfort were analyzed.

3.1.2 Methodology

3.1.2.1 Survey site

The size of the station determines whether it has sufficient capacity to carry customer flow. The indexes for measuring its size include the number of platforms, number of trains per day, and the number of dispatched passengers yearly[34]. Table 3.1 shows the sizes of the busiest top 20 railway stations in China. For this study, a large railway station with a length of 310m, a width of 190m, and a volume of $1.2 \times 10^6 m^3$ was chosen as the case study site. The selected station is representative: it is neither the largest nor the smallest among China's large railway stations; it has 18 platforms; and the number of passengers dispatched yearly is 110 million.

Case studies of passenger activities are common in China and most Asian countries and even in some European countries[35-37]. The mixed functions inside the selected case site are representative, commonly found in urban transit spaces[38,39], and include cafés, bars, restaurants, shops, security checks, ticket checks, information boards, pharmacies, and bathrooms. Almost all these functions are concentrated in one large space, and because the station is the main city transportation hub, the daily flow of people is very large, which indicates that the building is likely to have a complex acoustic environment[40].

To facilitate a high pedestrian movement rate throughout the station, a total of 18 escalators and 10 elevators provide full stairs-free access to all areas. In total, there are 570 seats within the station, and 20,000 people use the station every day (Table 3.2).

Table 3.1 The scale of railway stations in China (The busiest top 20)[34]

Name	Number of platforms	Number of trains/day	Number of passengers dispatched yearly (million)
Shanghai Hongqiao Railway Station	30	520	678
Guangzhou South Railway Station	28	653	470
Xi'an North Railway Station	34	186	111
Zhengzhou East Railway Station	30	352	135
Kunming South Railway Station	30	118	54.6
Nanjing South Railway Station	28	508	236
Hangzhou East Railway Station	28	243	51.8
Chengdu East Railway Station	26	593	293
Beijing South Railway Station	24	406	320
Changsha South Railway Station	24	399	168
Nanning East Railway Station	24	204	115
Shenzhen North Railway Station	20	258	98.2
Tianjin West Railway Station	18	236	217
Lanzhou West Railway Station	18	186	60.8
Shenyang South Railway Station	18	296	237
Harbin West Railway Station	18	225	110
Jin'an East Railway Station	18	259	146
Dalian North Railway Station	18	214	43.8
Taiyuan South Railway Station	18	209	48.6
Beijing West Railway Station	18	188	50.3

Chapter 3 Sound Environment and Acoustic Perception in Railway Stations

Table 3.2 Basic information on six typical spaces

		Seating area	Security check	Ticket check	Ticket lobby	Restaurant	Shop
Space type		Large space	Large space	Large space	Atrium space	Small space	Small space without a ceiling
Volume		11,100	180		864	172	288
Geometry (length/width)		162/68	36/5		36/24	21.5/8	24/12
Average customers		592	106	228	194	68	16
Photograph							
Interior materials and sound absorption coefficients	Ceilings	Gypsum $\alpha=0.3$	Gypsum $\alpha=0.3$	Gypsum $\alpha=0.3$	Gypsum $\alpha=0.3$	Gypsum $\alpha=0.3$	Gypsum $\alpha=0.3$
	Walls	Marble $\alpha=0.01$	Marble $\alpha=0.01$	Marble $\alpha=0.01$	Marble $\alpha=0.01$	Glass $\alpha=0.18$	Glass $\alpha=0.18$
	Floors	Marble $\alpha=0.01$	Marble $\alpha=0.01$	Marble $\alpha=0.01$	Marble $\alpha=0.01$	Ceramic $\alpha=0.02$	Marble $\alpha=0.01$
Sound absorber/reflector		Seat	X-ray security check machine, glass partition	Fare gate	Ticket machine/ window	Seat Table	Partition wall
Broadcast		With a broadcast	With a broadcast	With a broadcast	With a broadcast	Without a broadcast	With a broadcast
Behavioral patterns		Talking, resting	Talking, security checks	Ticket checks, talking, walking	Talking, ticket machine use, walking	Dining, walking, talking	Talking, walking

3.1.2.2 SPL and RT measurements

Previous studies have suggested that different sound sources and behavioral patterns influence the sound environment and the acoustic perceptions of users in open and indoor spaces and that the sound environment can, in turn, influence peoples' acoustic perceptions. The most important indexes that affect the sound environment are SPL and RT[13], which were measured by the following methods. The

119

measurement points are shown in Fig. 3.1. The selected test points covered six different functional spaces, including the seating area, security check, ticket lobby, ticket check, restaurants, and shops. The ticket lobby is located on both sides of the entrance and connected by a hallway to form a coupled space. The ticket window is on one side of the lobby; the rest of the lobby is typically full of people waiting to buy tickets. The other five spaces are all located in the waiting hall; the security checkpoint faces the entrance; there are four baggage screening machines; people queue through security to enter the waiting hall; the seating area is behind the security checkpoint; seats are divided into north and south banks, and each bank consists of 56 rows; the shops are located near the seating areas in the waiting hall; the outside is enclosed by a 2m high glass wall into a semi-open space; the ticket check is outside the seating area at the edge of the waiting hall; there is a large population density at check-in time; and the restaurants are small rooms on either side of the waiting hall.

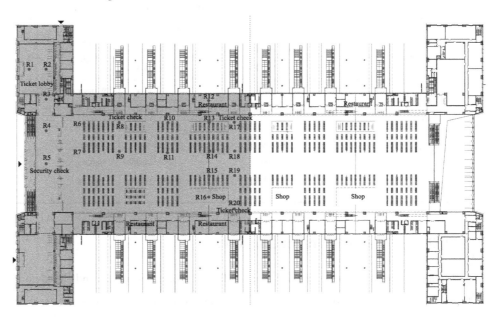

Fig. 3.1　The survey site and measurement points

Measurements were made during dense traffic periods. For each measurement point, a SPL meter was set to slow mode and A-weighting, and an instantaneous

Chapter 3 Sound Environment and Acoustic Perception in Railway Stations

reading was taken every 10s. To avoid sound source variability, each sound pressure level at each measurement point was tested 10 times; each measuring point was tested every hour, and the average value of the 10 sets of data was taken as the result of this measurement point. The measuring period lasted from 8:00 to 18:00. The equipment selection and measurement process followed the ISO 3382 standard. The sound level meter microphone was positioned 1m away from the walls and other main reflectors and 1.2-1.5m off the ground[41]. A total of 5 min of data were obtained at each measurement position, and the corresponding A-weighted equivalent SPL (LAeq) was derived. To avoid measurement error, each measurement in each space was taken from at least five random points at least 3m apart. To avoid the impact of speech on the measurements, no people were talking within 3m of the sound level meter[42]. The A-weighted SPLs measured at each point were averaged.

To understand the characteristics of the sound field in the waiting hall, the reverberation time in 6 areas of the waiting hall was measured at night when the trains had stopped. Only survey crews were present in the waiting hall during these measurements. An OS002 12-sided nondirectional sound source was adopted to play white noise[26]. As shown in Fig. 3.1, R denotes the position of a receiving point. An 801 sound level meter was used to test the reverberation time. Due to the large volume of the waiting hall, the difference in sound pressure level between most measuring point noise and background noise was less than the test range of T30. Therefore, the reverberation time of each area was compared with the T20 value.

3.1.2.3 Acoustic comfort survey

Acoustic comfort is a key evaluation index for the soundscapes of indoor spaces[43]. Thus, this study examined the influences of different spaces in the evaluation of passenger acoustic comfort using a questionnaire survey method. In this study, a total of 300 questionnaires were issued, 50 in each representative space, among which 289 were valid. The participants were of different ages and a balanced male-to-female ratio was maintained: the participants were between 15 and 80 years old, and the male-to-female ratio was set to 1.02 : 1 (146 males and 143 females) to ensure that the sample sex ratio was balanced.

To ensure the representativeness of the selected spaces, a preliminary survey of six typical spaces in the station was conducted before the formal investigation[30]. The contents of the investigation concerned sound sources, personnel behavior, sound field characteristics and comfort evaluation. The results obtained from the six spaces were typical and obviously diverse. Previous studies have also indicated that an interview duration of more than 5min may decrease the reliability of the investigation[44]; therefore, the questionnaires in this study were all delivered and completed within 2-3min. Approximately 10 interviews were conducted at each survey point. Participants were interviewed individually and briefed on the purpose of the study; then, they provided written informed consent to participate in the research. The survey points are marked with solid circles in Fig. 3.1, and Table 3.3 shows the questions, which included four social factors (Nos. 1-6) and a subjective evaluation (Nos. 7-14). Previous studies have shown that social factors may cause different evaluation results[45], and therefore, Nos. 1-4 provide a survey of the social background. When people arrive at the waiting hall at different time periods, they may give different evaluations of the comfort level of the acoustic environment. In addition, the time people spend in the space may also make a difference in their evaluation of environmental comfort. Nos. 5-6 are intended to address the above questions. Nos. 7-10 provide a subjective evaluation of the total sound environment and Nos. 11-15 provide a subjective evaluation of each sound source. No. 7 is an overall sound environment evaluation; No. 8 provides an evaluation of the comfort level; No. 9 asks the participants to evaluate the language intelligibility in the overall sound environment and thereby indirectly evaluate the level of background noise; and No. 10 provides a subjective evaluation of the reverberation time. The existing research showed that the acoustic comfort of sound sources is related to sound characteristics such as loudness, intelligibility, noise level and preference degree, as shown in Table 3.3, Nos. 12-15. Loudness is a subjective measurement describing the strength of the ear's perception of a sound[46]. Intelligibility is a measure of speech comprehensibility during communication[47]. Sound level refers to various logarithmic measurements of audible vibrations[48], and preference degree is related to a person's degree of preference for one or more sound sources[49].

Table 3.3 Questionnaire questions and scales

No.	Questions	Scales
1	Gender	1, male; 2, female
2	Age	1, <20; 2, 20-40; 3, 41-60; 4, >60
3	Education level	1, primary; 2, secondary; 3, higher education
4	Income	1, <1,000; 2, 1,000-2,000; 3, 2,001-3,000; 4, 3,001-4,000; 5, 4,001-5,000; 6, >5,000
5	Visit time	1, morning (9:00-11:59); 2, midday (12:00-14:59); 3, afternoon (15:00-17:59); 4, evening (18:00-21:00)
6	Visit duration	1, less than 1h; 2, 1-2h; 3, more than 2h
7	Evaluation of the overall sound environment	Scale of 1 to 5, with 1 being very noisy and 5 being very quiet
8	Acoustic comfort of the overall sound environment	Scale of 1 to 5, with 1 being very uncomfortable and 5 being very comfortable
9	Sound volume of communicating with companions	Scale of 1 to 5, with 1 being very loud and 5 being very soft
10	Subjective impression of reverberation	Scale of 1 to 5, with 1 being very long and 5 being very short
11	Acoustic comfort of various sound sources	Scale of 1 to 5, with 1 being very uncomfortable and 5 being very comfortable
12	Loudness of various sound sources	Scale of 1 to 5, with 1 being very low and 5 being very high
13	Intelligibility of various sound sources	Scale of 1 to 5, with 1 being very clear and 5 being very unclear
14	Noise level of various sound sources	Scale of 1 to 5, with 1 being very noisy and 5 being very quiet
15	Preference degree of various sound sources	Scale of 1 to 5, with 1 being highly disliked and 5 being highly liked

The attitudes of participants were measured using a Likert scale, which has been widely used in survey research of environmental effects on subjective comfort[50, 51]. Regarding acoustic comfort, the interviewees provided answers using the following five-point Likert-type scale: 1, very uncomfortable; 2, uncomfortable; 3, neither comfortable nor uncomfortable; 4, comfortable; and 5, very comfortable[52]. The reliability coefficient of the questionnaire was estimated as 0.82 (Cronbach's alpha). The Kaiser-Meyer-Olkin (KMO) values of the subscales were greater than 0.5. For the Bartlett spherical test, $p < 0.01$, with a reliability coefficient of $0.9 > \alpha \geqslant 0.8$, indicating that the questionnaire satisfied the reliability requirement[53].

3.1.2.4 Data statistics and analysis

The results of the subjective and objective measurements were analyzed using SPSS 15.0 software[54]. Pearson's correlation coefficient was used to determine the factors and dominant sound sources that affected people's comfort evaluations of the sound environment, and mean differences (t-tests, two-tailed) were used to investigate the influences of the existence or nonexistence of dominant background sound sources on the participants. Pearson's correlation and regression analysis were used to determine the factors affecting the acoustic comfort of the dominant sound sources from the sound source characteristics. The factors affecting people's acoustic comfort evaluation are discussed from the perspectives of demographic and social factors.

3.1.3 Results and analysis

Based on the survey and measurement results, this section discusses the following effects: different types of sound sources, dominant sound sources, demographic and social factors on sound level, sound perception, loudness and acoustic comfort.

3.1.3.1 Overall comfort level and sound environment

Fig. 3.2 shows the subjective evaluations of the overall sound environment in the six spaces and includes the mean and standard deviation of each evaluation. The comfort of the sound environment in the railway station was acceptable (mean value of 3.65). However, the comfort evaluations in the seating area and shop were relatively higher (mean values of 3.81 and 3.91, respectively), and the comfort evaluations in the restaurant were slightly lower (mean value of 3.28). Sound level and speech intelligibility were considered the most important influencing factors in the sound environment[55]. The SPL and RT measurement results are shown in Table 3.4. The seating area, shop area and ticket lobby were quieter, and the areas with high concentrations of people are noisy. RT is related to the size of the space: the larger the space is, the longer the RT is.

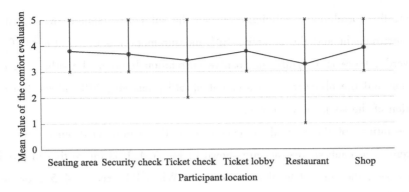

Fig. 3. 2　Means and standard deviations of the comfort evaluations of the overall sound environment

Table 3. 4　Measurement results for each measurement point in the six areas

Space	Ticket lobby			Security check		Restaurant		Shop
Measurement point	R1	R2	R3	R4	R5	R12		R16
Leq [dB(A)]	68.2	67.7	68.4	69.8	70.8	75.1		58.7
RT (s)	5.16	4.98	5.39	2.97	3.15	1.58		4.62
Space	Ticket check							
Measurement point	R8	R13	R17	R20				
Leq [dB(A)]	72.8	68.9	73.8	73.2				
RT (s)	3.28	4.16	3.59	3.36				
Space	Seating area							
Measurement point	R6	R7	R9	R10	R11	R14	R15	R18	R19
Leq [dB(A)]	64.9	62.1	66.4	65.2	60.8	64.2	62.6	66.0	65.8
RT (s)	8.96	9.64	9.69	9.98	8.63	9.15	9.38	8.79	8.91

Fig. 3. 3 illustrates the individual response ratings regarding the overall environment vs. the SPL measurement at each survey site. The overall trend is that the comfort level and sound volume of communication decrease with increasing SPL. However, it is interesting to note that in addition to the measuring point of the restaurant, the other measuring points also show reduced comfort levels as the sound pressure level increases. The average SPL values in the seating area, ticket lobby and shops were all below 70dB. The restaurants are the noisiest places, with an average SPL value of 75.1dB, but the comfort level and the appropriateness rating in the restaurants are higher than those in the ticket check area. According to the Pearson correlation

analysis, the correlation coefficient between the subjective comfort evaluation of the sound environment and the objective SPL measurement was 0.513 ($p < 0.01$). In other words, there is a significant positive correlation between the subjective comfort evaluation and the objective measurement of SPL, namely, SPL affects the comfort evaluation of the sound environment.

The ratings of the individual responses on the overall environment vs. the RT measurement at each survey site are shown in Fig. 3.4. It is interesting to see that as RT increases, the comfort level also increases. When RT exceeds 4.5s, participants can feel the reverberation in the space. High RT also increases the communication sound volume because high RT increases the background noise and reduces speech intelligibility[56].

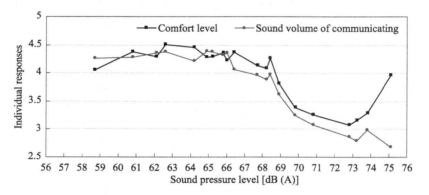

Fig. 3.3 Ratings of the comfort level and appropriateness with mean values of the SPL measurements

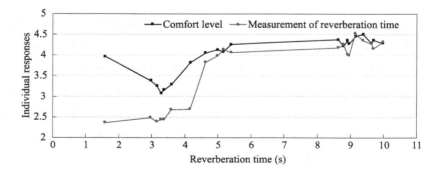

Fig. 3.4 Ratings of the comfort level and subjective impression of reverberation with the RT measurements

3.1.3.2 Sonic composition

Participants were asked to list three sounds that they heard at that moment and provide a comfort scale (scale of 1 to 5, with 1 being very uncomfortable and 5 being very comfortable) to identify various background noise sound sources and determine the types of the sound sources from participants' perspective. Sound sources mentioned fewer than five times were ignored[57]. Finally, the various individual sound sources in railway stations were divided into five types: broadcasts, speech sounds, activity sounds, mechanical noise and luggage noise. The key sounds and comfort scale in each space are shown in Fig. 3.5. Interestingly, the participants gave lower evaluations of the comfort level at the survey points in spaces with high-density crowds, such as ticket checks and restaurants, and they were significantly influenced by activity sounds and speech sounds. The participants found the spaces around machines to be a generally poor acoustic environment, and it appeared that people were bothered more by mechanical noise and luggage noise. In particular, as a key sound, most participants gave a high appropriateness score for broadcasts, and people interacting and communicating in spaces were not significantly annoyed by broadcasts.

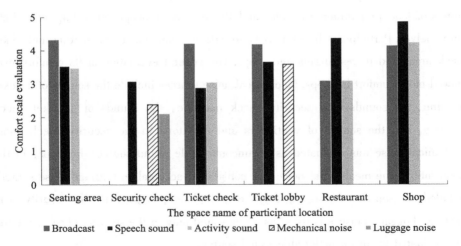

Fig. 3.5 **Key sounds and mean comfort evaluation in different spaces**

Participants were also asked to list five sound sources that they heard in each space and provide their evaluation of the acoustic comfort. Table 3.5 (Column A) shows the means and standard deviations of the participants' acoustic comfort evaluation of various background noise sound sources in the spaces. As a key sound, broadcasts were fundamental in most spaces. This sound did not garner much attention in the security check, but participants listed broadcasts as a key sound source in the other spaces, and the acoustic comfort evaluation of broadcasts tended to be given a comfortable score. The speech sound sources consisted of the sounds of participants (the speech sounds of companions and other participants, shouting, crying and phone calls of people moving around) and the speech sounds of staff. Speech sounds were mentioned as key sounds in every space, and the participants tended to provide a better evaluation of the comfort level of speech sounds in commercial spaces, including restaurants and shops. The comfort level of speech sound was the lowest in the ticket check area and was evaluated as uncomfortable. A comparison of the measurement results of SPL found that SPLs of the test points R8, R17 and R20 near the ticket check all exceeded 70dB, indicating that when SPL is above a certain level enhances the annoyance degree of speech sound, resulting in low comfort-level evaluations in this area. Activity sound sources were caused by user activities in these spaces, including impact sounds from footsteps, scratching sounds from clothes, the sounds of food preparation by staff, and the sounds of people choosing goods from store shelves. Participants listed activity sounds as dominant in the seating and ticket check areas and in restaurants and shops. The general evaluation of its comfort level showed more comfort in shops. Mechanical noise sources include the sounds of a ticket machine, the sounds of a security check machine, the sounds of a ticket check machine, and the sounds of ventilators and elevators. In the security check area, mechanical noise was evaluated as an uncomfortable sound source; however, in the ticket lobby, the mechanical noise was considered acceptable. Luggage noise sources included the sounds of dragging luggage and placing luggage in the security scan machine. Luggage noise was listed as a key sound only in the security check area and was evaluated as an uncomfortable sound source.

Table 3.5 provides a statistical analysis using the Pearson correlation of the acoustic comfort evaluations of various individual sound sources and the comfort evaluation of the overall sound environment in each space ($p < 0.01$). The results showed that positive correlations among the following: acoustic comfort evaluations of broadcast sounds, the speech sounds of other participants, and shouting in the seating area; the speech sounds of the staff, the security check machine and the sound of placing luggage in the security scan machine in security check area; broadcast sounds, conversational sounds from other participants and the sound of the ticket machine in the ticket lobby; the speech sounds of staff and the food preparation by staff in the restaurant; and the broadcast and the chatting sounds of other participants in the shops. The correlation coefficients ranged from 0.25-0.5. To determine the influences of these sound sources on the overall comfort of the sound environment, an independent samples t-test was conducted in both the presence and absence of the sound sources. The results (Table 3.5, Column C) showed that the comfort evaluations of the overall sound environment in the presence or absence of shouting in the seating area, luggage noise in the security check, the speech sounds of staff at the ticket check and shouting in the ticket lobby all displayed marked differences. The comfort evaluation (mean of 2.96) of the overall sound environment in the seating area in the presence of shouting was significantly lower than that (mean of 3.42) in the absence of shouting; the comfort evaluation (mean of 2.11) of the overall sound environment in the security check in the presence of luggage noise was lower than that (mean of 2.69) in the absence of luggage noise; the comfort evaluation (mean of 2.49) of the overall sound environment in the ticket check in the presence of the speech sounds of staff was lower than that (mean of 2.92) in the absence of the speech sounds of staff; and the comfort evaluation (mean of 3.01) of the overall sound environment in the ticket lobby in the presence of shouting was lower than that (mean of 3.42) in the absence of shouting. The presence or absence of broadcast and chatting sounds of other participants in the seating area; the speech sounds of staff and security check machines in the security check; the broadcast and chatting sounds of other participants in the ticket check; the broadcast and ticket machine noise in the ticket lobby; the speech sounds of staff and food preparation by staff in restaurants; and the broadcast and chatting sounds of other participants in shops showed no significant effect on the comfort evaluation of the overall sound environment.

Table 3.5 Correlation analysis between the acoustic comfort of various sound sources and the overall sound environment comfort evaluation

Name of space	Type of sound source (Only include key sound sources)		A	B	C
Seating area	Broadcast	Broadcast information	4.31/0.68	0.293/0.000 (**)	$p=0.072>0.05$
	Speech sounds	Speech sounds of companions	4.02/0.81	0.223/0.021	—
		Chatting sounds of other people	3.97/1.05	0.492/0.000 (**)	$p=0.075>0.05$
		Shouting	2.96/0.80	0.228/0.000 (**)	$p=0.035<0.05$
		Crying	3.08/0.89	0.245/0.037	—
Security check	Speech sounds	Speech sounds of companions	3.56/0.86	0.198/0.026	—
		Chatting sounds of other people	3.01/0.88	0.312/0.053	—
		Speech sounds of staff	2.86/0.89	0.332/0.000 (**)	$p=0.081>0.05$
	Mechanical noise	Security check machines	2.39/0.93	0.255/0.000 (**)	$p=0.093>0.05$
	Luggage noise	Placement of luggage	2.11/0.69	0.366/0.000 (**)	$p=0.035<0.05$
Ticket check	Broadcast	Broadcast information	4.18/1.02	0.258/0.000 (**)	$p=0.065>0.05$
	Speech sounds	Speech sounds of companions	3.29/0.87	0.218/0.041	—
		Chatting sounds of other people	2.98/0.82	0.308/0.000 (**)	$p=0.072>0.05$
		Speech sounds of staff	2.49/0.79	0.281/0.000 (**)	$p=0.035<0.05$
		Shouting	2.36/0.83	0.186/0.044	—
Ticket lobby	Broadcast	Broadcast information	4.16/0.92	0.292/0.000 (**)	$p=0.064>0.05$
	Speech sounds	Speech sounds of companions	3.87/0.83	0.124/0.128	—
		Chatting sounds of other people	3.49/0.69	0.146/0.013	—
		Shouting	3.01/0.78	0.322/0.000 (**)	$p=0.035<0.05$
	Mechanical noise	Ticket machines	3.56/0.96	0.251/0.000 (**)	$p=0.082>0.05$
Restaurant	Broadcast	Broadcast information	3.08/0.85	0.243/0.029	—
	Speech sounds	Speech sounds of companions	4.47/0.92	0.242/0.086	—
		Chatting sounds of other people	4.27/1.01	0.262/0.069	—
		Speech sounds of staff	4.36/1.14	0.265/0.000 (**)	$p=0.095>0.05$
	Activity sounds	Food preparation by staff	3.08/0.96	0.278/0.000 (**)	$p=0.068>0.05$

Continued

Name of space	Type of sound source (Only include key sound sources)		A	B	C
Shop	Broadcast	Broadcast information	4.12/0.98	0.226/0.000 (**)	$p=0.089>0.05$
	Speech sounds	Speech sounds of companions	4.95/0.82	0.206/0.028	—
		Chatting sounds of other people	4.78/0.99	0.288/0.000 (**)	$p=0.091>0.05$
		Speech sounds of staff	4.82/0.83	0.186/0.058	—
	Activity sounds	Choosing goods from store shelves	4.22/0.79	0.229/0.083	—

Note: A. Mean and standard deviation of the acoustic comfort evaluations of the sound sources.

B. Correlation coefficient and significance level of acoustic comfort evaluation of various sound sources and the overall sound environment comfort evaluation.

C. p-value of independent samples t-test.

** indicates $p<0.01$.

3.1.3.3 Sound characteristics of dominant individual sound sources

In this section, the subjective evaluations of the loudness, intelligibility, noise level and preference degree based on the questionnaire survey are analyzed. Fig. 3.6 summarizes the sound source characteristics by averaging the scores of four factors for different sound sources. Broadcasts, which drew people's attention, exhibited the lowest sound level, resulting in a high preference degree (with a mean value of 4.21), although the loudness and intelligibility were high. However, there were variations in the luggage noise and mechanical noise; they had a high sound level and loudness but a low preference degree, meaning that these two sound sources resulted in a high annoyance degree (discomfort) of people (means of 2.65 and 2.54, respectively). The four sound characteristics of speech and activity sounds were moderate, with mean values of 3-4.

Because the evaluation of the sound characteristics of the dominant sound sources may have different influences on SPL and acoustic comfort, SPLs and acoustic comfort of different sound source characteristics from different dominant sound sources were also compared. Mechanical noise and luggage noise are not analyzed in this section due to a

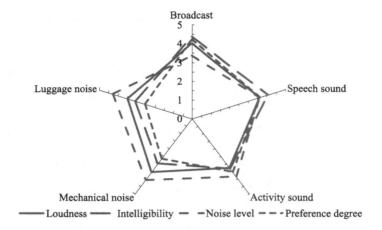

Fig. 3.6 Sound source characteristics

lack of examples. The relationships between the measured LAeq and sound source characteristics as well as acoustic comfort and sound source characteristics with different sound sources are shown in Fig. 3.7 and Fig. 3.8, where the linear regressions and the coefficients of determination (R^2) are also presented. There is a general correlation between the measured sound level and loudness and intelligibility as well as with the sound level for some of the sound sources ($p < 0.001$).

As the measured overall sound level increases, the sound level of speech also increased; the R^2 value was 0.515. Other sound sources showed only weak correlations. A possible reason for this difference is that broadcasts and activity sounds were not always present, but speech sounds were always present as background noise, and noise from speech was the most important factor affecting the sound environment. It is interesting to note that as the measured sound level increased, the intelligibility of both the broadcast and speech sound also decreased; the R^2 values were 0.511 (broadcast) and 0.532 (speech sound). A possible reason is that the increase in SPL is mainly caused by these two sounds; they interfere with each other such that one voice obscures the other, decreasing the intelligibility of both. It is also interesting to note that as the measured sound level increased, the loudness of three sound sources decreased; sound masking among these sound sources may have led to this result. There was a significant correlation ($R^2 = 0.791$) between acoustic comfort and the speech sound level, indicating that noise from speech sounds is not generally liked.

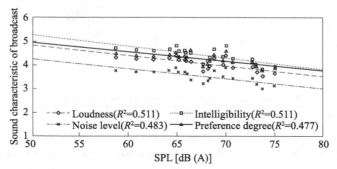

(a) Relationship between SPL and the sound characteristics of broadcast

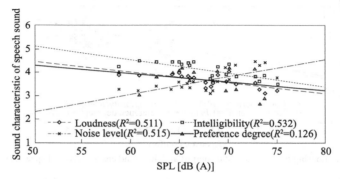

(b) Relationship between SPL and the sound characteristics of speech sounds

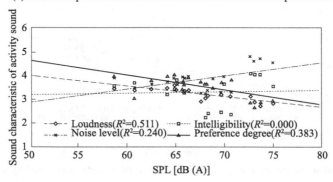

(c) Relationship between SPL and the sound characteristics of activity sounds

Fig. 3.7 The relationship between SPL and the sound characteristics of the dominant sound sources

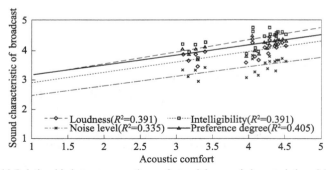

(a) Relationship between acoustic comfort and the sound characteristics of broadcast

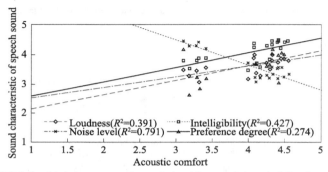

(b) Relationship between acoustic comfort and the sound characteristic of speech sounds

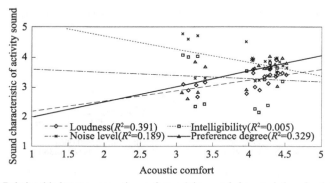

(c) Relationship between acoustic comfort and the sound characteristics of activity sounds

Fig. 3. 8   The relationship between acoustic comfort and the sound characteristics of the dominant sound sources

3.1.3.4 Effects of demographic and social factors

The mean difference between males and females in the evaluation of acoustic comfort was determined in every space. As shown in Table 3.6, no significant differences ($p < 0.1$) were found between males and females. These results were consistent with those of previous studies, which suggested that the effect of gender on sound annoyance evaluation is generally insignificant (Meng and Kang, 2016). However, age difference was significant ($p < 0.01$ or $p < 0.05$); the correlation coefficients ranged from 0.16 to 0.28 in most spaces. Acoustic comfort was higher for older people, and our results are consistent with those of a previous study in Europe[1], which showed that cultural aspects were not the main factor influencing older people's noise perceptions. The same conclusion has been drawn for other types of buildings[58]. Education level and income difference were also significant factors ($p < 0.01$ or $p < 0.05$) in people's acoustic comfort; the correlation coefficients ranged from 0.13 to 0.22 for education level and 0.15 to 0.22 for income for the tested spaces. It is interesting to note that acoustic comfort increased with higher education level and income in quiet places (where the measurement results of SPL were below 70dB), while it usually decreased with higher education level and income in noisy places (where the measurement results of SPL exceeded 70dB). This result indicates that people with different income levels have different tolerances for different SPLs. Differences in visit frequency were associated with a significant difference in the comfort evaluation of the sound environment in six spaces ($p < 0.05$): people who visited the station frequently (mean = 3.46) gave a more critical evaluation than people who did not (mean = 3.12). It was also found that the visit duration in space was significantly correlated ($p < 0.01$) with acoustic comfort, and under a high SPL, visit duration and acoustic comfort had a significant negative correlation. This indicates that people tend to be more annoyed when they spend more time in a high-SPL environment.

Table 3.6 The relationship between acoustic comfort and participant demographic information

Social factors	Seating area	Security check	Ticket check	Ticket lobby	Restaurant	Shop
Sex	0.03	0.03	0.01	0.04	0.01	0.02
Age	0.28 **	0.16 *	0.18 *	0.20 **	0.15 *	0.08
Education level	0.18 *	0.16 *	-0.22 **	0.15 *	-0.13 *	0.21 **
Income	0.22 **	-0.19 **	-0.25 **	0.15 *	0.06	0.24 **
Visit time	0.18 *	0.21 **	-0.26 **	0.20 **	0.15 *	0.29 **
Visit duration	0.15 **	0.28 **	-0.31 **	0.16 **	0.11 **	0.16 **

Note: ** indicates $p < 0.01$ and * indicates $p < 0.05$.

3.1.4 Conclusions

Based on measurements and the results of a questionnaire survey conducted at a railway station, the effects of various sound sources on the sound environment and the corresponding subjective evaluations of acoustic comfort were studied.

With regard to the overall sound environment and comfort level, according to the measurement and survey results, SPLs were concentrated between 60-70dB (A). There was a significant positive correlation between the subjective comfort evaluation and sound level measurement: the correlation coefficient was 0.513 ($p < 0.01$). The comfort level and appropriateness were moderately negatively correlated with SPL at the station. It is interesting to see that people's comfort level increases with increasing RT, which is contrary to popular belief. This result occurs because the test points with lower SPL values have longer RT, and the comfort level is more highly affected by SPL, leading to a higher comfort rating. The RT was difficult for people to sense; when it exceeded 4.5s, the participants could feel the reverberation in the space. High RT also increases the sound volume during communication.

The sonic composition of sound sources in the railway station included broadcasts, speech sounds, activity sounds, mechanical noise and luggage noise. The dominant sound sources differed in each space. Broadcasts, speech sounds, and activity sounds were the sounds most identified by the participants. The acoustic comfort of broadcasts was the most preferable; speech sounds were preferred in restaurants and shops but considering annoying in security and ticket check areas, and the acoustic comfort of mechanical noise and luggage noise was the least preferable in security check areas. Among the sound sources, the speech sounds of other people in the seating area, the speech sounds of staff, the sounds of placing luggage in the security scan machine in the security check area, the sound of the ticket machine in the ticket lobby, and the speech sounds of staff in the restaurants significantly influenced the participants' acoustic comfort evaluations.

With regard to the effects of various sound sources on the comfort level, the sound characteristics of five main sound sources were analyzed. The results show that people have the highest preference for broadcasts, followed by speech sounds and activity sounds. The intelligibility of broadcasts is low because they are easily obscured by other sound sources, such as speech sounds and activity sounds. However, people want to hear the broadcasts; therefore, the intelligibility of the broadcasts needs to be improved. Although the participants' preferences for speech sound and activity sound were at high levels, as the density of people increases, the preferences for those types of sound decline rapidly. Therefore, those two sound source types need to be controlled, especially in large spaces with high population densities. People dislike luggage and mechanical sounds because of their high sound levels and perceived loudness; these two sound sources received considerable attention and need to be reduced. To improve the acoustic environment, sound sources with both low preferences and high loudness were considered. It is worth investigating the sound environment of large railway stations from a soundscape perspective to determine whether measures might be implemented that would enhance users' acoustic comfort.

With regard to demographic and social factors, age differences resulted in significant differences in comfort evaluations: older people provided higher comfort evaluations. People's education level and income differences also resulted in significant differences ($p < 0.01$ or $p < 0.05$) in acoustic comfort: higher incomes and education levels were associated with high evaluations of acoustic comfort in quiet places but low evaluations in noisy places. Both visit frequency and visit duration were significantly different: people who visited the station frequently gave a more critical evaluation, and visit duration and acoustic comfort had a significantly negative correlation.

While in this study the station spaces are rather large, it is also interesting to consider other space types such as long spaces in further studies[59-62].

References

[1] WU Y, KANG J. The complexity of sound environment contributing to acoustic comfort in urban intermodal transit spaces [J]. INTER-NOISE and NOISE-CON Congress and Conference Proceedings, Inter-noise 2019, Madrid, Spain: 257-264.

[2] WESTON A. The Soundscape: our sonic environment and the tuning of the world [J]. Environmental Ethics, 1996, 18 (3): 331-333.

[3] HUANG L, ZHU Y, OUYANG Q, et al. A study on the effects of thermal, luminous, and acoustic environments on indoor environmental comfort in offices [J/OL]. Building and Environment, 2012, 49: 304-309. https://doi.org/10.1016/j.buildenv.2011.07.022.

[4] CHEN X, KANG J. Acoustic comfort in large dining spaces [J/OL]. Applied Acoustics, 2017, 115: 166-172. https://doi.org/10.1016/j.apacoust.2016.08.030.

[5] XIAO J, ALETTA F. A soundscape approach to exploring design strategies for acoustic comfort in modern public libraries: a case study of the Library of Birmingham [J]. Noise Mapping, 2016, 3 (1): 264-273.

[6] MENG Q, JIN H, KANG J. Prediction of subjective loudness in underground shopping streets using artificial neural networks [J/OL]. Noise Control Engineering Journal, 2012, 60 (3): 329-339. https://doi.org/10.3397/1.3701010.

[7] CASSINA L, FREDIANELLI L, MENICHINI I, et al. Audio-visual preferences and tranquillity ratings in urban areas [J/OL]. Environments, 2018, 5 (1): 1. https://doi.org/10.3390/environments5010001.

[8] DE COENSEL B, BOTTELDOOREN D. The quiet rural soundscape and how to characterize it [J]. Acta Acustica united with Acustica, 2006, 92 (6): 887-897.

[9] RUIZ-PADILLO A, RUIZ D P, TORIJA A J, et al. Selection of suitable alternatives to reduce the environmental impact of road traffic noise using a fuzzy multi-criteria decision model [J/OL]. Environmental Impact Assessment Review, 2016, 61: 8-18. https://doi.org/10.1016/j.eiar.2016.06.003.

[10] LICITRA G, FREDIANELLI L, PETRI D, et al. Annoyance evaluation due to overall railway noise and vibration in Pisa urban areas [J/OL]. Science of the total environment, 2016, 568: 1315-1325. https://doi.org/10.1016/j.scitotenv.2015.11.071.

[11] GAGLIARDI P, FREDIANELLI L, SIMONETTI D, et al. ADS-B system as a useful tool for testing and redrawing noise management strategies at Pisa Airport [J/OL]. Acta Acustica united with Acustica, 2017, 103 (4): 543-551. https://doi.org/10.3813/AAA.919083.

[12] BERNARDINI M, FREDIANELLI L, FIDECARO F, et al. Noise assessment of small vessels for action planning in Canal Cities [J/OL]. Environments, 2019, 6 (3): 31. https://doi.org/10.3390/environments6030031.

[13] TAVOSSI H M. Traffic noise attenuation by scattering, resonance and dispersion [J/OL]. The Journal of the Acoustical Society of America, 2003, 114 (4): 2353-2353. https://doi.org/10.1121/1.4781158.

[14] MAKAREWICZ R, ZÓLTOWSKI M. Variations of road traffic noise in residential areas [J/OL]. The Journal of the Acoustical Society of America, 2008, 124 (6): 3568-3575. https://doi.org/10.1121/1.3008003.

[15] WILLIAMS I D, MCCRAE I S. Road traffic nuisance in residential and commercial areas [J/OL]. Science of the total environment, 1995, 169 (1-3): 75-82. https://doi.org/10.1016/0048-9697 (95) 04635-E.

[16] STANSFELD S A, BERGLUND B, CLARK C, et al. Aircraft and road traffic noise and children's cognition and health: a cross-national study [J/OL]. The Lancet, 2005, 365 (9475): 1942-1949. https://doi.org/10.1016/S0140-6736 (05) 66660-3.

[17] IGLESIAS-MERCHAN C, DIAZ-BALTEIRO L, SOLIÑO M. Transportation planning and quiet natural areas preservation: aircraft overflights noise assessment in a national park [J/OL]. Transportation research part D: transport and environment, 2015, 41: 1-12. https://doi.org/10.1016/j.trd.2015.09.006.

[18] CASAZZA M, BOGGIA F, SERAFINO G, et al. Environmental impact assessment of an urban port: noise pollution survey in the port area of Napoli (S Italy) [J/OL]. Journal of Environmental Accounting and Management, 2018, 6 (2): 125-133. DOI: 10.5890/JEAM.2018.06.004.

[19] BABISCH W, HOUTHUIJS D, PERSHAGEN G, et al. Associations between road traffic noise level, road traffic noise annoyance and high blood pressure in the HYENA study [J]. Journal of the Acoustical Society of America, 2008, 123 (5): 3448-3448.

[20] HATANO M. Noise impact of rail passenger service [C] //INTER-NOISE and NOISE-CON Congress and Conference Proceedings. Institute of Noise Control Engineering, 1982, 1982

(2): 201-204.

[21] FYHRI A, AASVANG G M. Noise, sleep and poor health: modeling the relationship between road traffic noise and cardiovascular problems [J/OL]. Science of the Total Environment, 2010, 408 (21): 4935-4942. https://doi.org/10.1016/j.scitotenv.2010.06.057.

[22] KLATTE M, MEIS M, SUKOWSKI H, et al. Effects of irrelevant speech and traffic noise on speech perception and cognitive performance in elementary school children [J]. Noise and Health, 2007, 9 (36): 64.

[23] MÜNZEL T, GORI T, BABISCH W, et al. Cardiovascular effects of environmental noise exposure [J/OL]. European Heart Journal, 2014, 35 (13): 829-836. https://doi.org/10.1093/eurheartj/ehu030.

[24] CLARK C, MARTIN R, VAN KEMPEN E, et al. Exposure-effect relations between aircraft and road traffic noise exposure at school and reading comprehension: the RANCH project [J/OL]. American Journal of Epidemiology, 2005, 163 (1): 27-37. https://doi.org/10.1093/aje/kwj001.

[25] STANSFELD S A, MATHESON M P. Noise pollution: non-auditory effects on health [J/OL]. British Medical Bulletin, 2003, 68 (1): 243-257. https://doi.org/10.1093/bmb/ldg033.

[26] WANG C, MA H, WU Y, et al. Characteristics and prediction of sound level in extra-large spaces [J/OL]. Applied Acoustics, 2018, 134: 1-7. https://doi.org/10.1016/j.apacoust.2017.12.023.

[27] MYERS T D, BALMER N J. The impact of crowd noise on officiating in Muay Thai: achieving external validity in an experimental setting [J]. Frontiers in Psychology, 2012, 3: 346.

[28] HAYNE M J, TAYLOR J C, RUMBLE R H, et al. Prediction of noise from small to medium sized crowds [J]. Proceedings of Acoustics 2011, 2011.

[29] NIE S S, KANG J. An acoustic model of crowd in large spaces [J]. Journal of Applied Acoustic, 2016, 35 (2): 128-136. (in Chinese).

[30] WU Y, KANG J, ZHENG W. Acoustic environment research of railway station in China [J/OL]. Energy Procedia, 2018, 153: 353-358. https://doi.org/10.1016/j.egypro.2018.10.z038.

[31] BANDYOPADHYAY P, BHATTACHARYA S K, KASHYAP S K. Assessment of noise environment in a major railway station in India [J/OL]. Industrial Health, 1994, 32 (3): 187-192. https://doi.org/10.2486/indhealth.32.187.

[32] LIU G, HOU D, WANG L, et al. Acoustics testing and simulation analysis of waiting hall

in the line-side high-speed railway station [J/OL]. The Journal of the Acoustical Society of America, 2014, 135 (4): 2332-2332. https://doi.org/10.1121/1.4877651.

[33] TAO Z, WANG Y, ZOU C, et al. Assessment of ventilation noise impact from metro depot with over-track platform structure on workers and nearby inhabitants [J]. Environmental Science and Pollution Research, 2019, 26 (9): 9203-9218.

[34] The ranking of China's top high-speed railway stations in 2018 based on big data [R/OL]. https://www.135995.com/207/163461.html; 2018 [accessed 04 August 2018].

[35] LI Z H, ZHANG X. Simulation and analysis based on emergency evacuation success rate in elevated layer of Beijing South Railway Station [M] //Applied mechanics and materials. Trans Tech Publications, 2014, 587: 1912-1915.

[36] SHAH J, JOSHI G J, PARIDA P. Behavioral characteristics of pedestrian flow on stairway at railway station [J/OL]. Procedia-social and behavioral sciences, 2013, 104: 688-697. https://doi.org/10.1016/j.sbspro.2013.11.163.

[37] PRASSLER E, SCHOLZ J, ELFES A. Tracking people in a railway station during rush-hour [M] //International Conference on Computer Vision Systems. Springer, Berlin, Heidelberg, 1999: 162-179.

[38] LIN Y H, CHEN C F. Passengers' shopping motivations and commercial activities at airports: the moderating effects of time pressure and impulse buying tendency [J/OL]. Tourism Management, 2013, 36: 426-434. https://doi.org/10.1016/j.tourman.2012.09.017.

[39] WANG Y, PENG J. The construction study of place sense of tourism in the areas of high-speed railway station and airport [J]. Tourism Research, 2014, 1.

[40] YUE W, JIAN K, SHIWEI D. A study on evacuation movement time in elevated railway waiting halls [J]. Architectural Journal, 2015 (1): 100-105.

[41] MENG Q, ZHANG S, KANG J. Effects of typical dining styles on conversation behaviours and acoustic perception in restaurants in China [J/OL]. Building and Environment, 2017, 121: 148-157. https://doi.org/10.1016/j.buildenv.2017.05.025.

[42] ZAHORIK P. Assessing auditory distance perception using virtual acoustics [J/OL]. The Journal of the Acoustical Society of America, 2002, 111 (4): 1832-1846. https://doi.org/10.1121/1.1458027.

[43] LONG M. Architectural acoustics [M]. Elsevier Academic Press. 2006.

[44] KANG J. Urban sound environment [M]. London: Taylor and Francis. 2006.

[45] MENG Q, KANG J. The influence of crowd density on the sound environment of commercial pedestrian streets [J/OL]. Science of the Total Environment, 2015, 511: 249-258. https://doi.org/10.1016/j.scitotenv.2014.12.060.

[46] ZWICKER E. Procedure for calculating loudness of temporally variable sounds [J/OL]. The Journal of the Acoustical Society of America, 1977, 62 (3): 675-682. https://doi.org/10.1121/1.381580.

[47] FRENCH N R, STEINBERG J C. Factors governing the intelligibility of speech sounds [J/OL]. The Journal of the Acoustical Society of America, 1947, 19 (1): 90-119. https://doi.org/10.1121/1.1916407.

[48] BURKARD R. Sound pressure level measurement and spectral analysis of brief acoustic transients [J/OL]. Electroencephalography and Clinical Neurophysiology, 1984, 57 (1): 83-91. https://doi.org/10.1016/0013-4694 (84) 90010-5.

[49] YU L, KANG J. Factors influencing the sound preference in urban open spaces [J/OL]. Applied Acoustics, 2010, 71 (7): 622-633. https://doi.org/10.1016/j.apacoust.2010.02.005.

[50] LIU F, KANG J. Relationship between street scale and subjective assessment of audio-visual environment comfort based on 3D virtual reality and dual-channel acoustic tests [J/OL]. Building and Environment, 2018, 129: 35-45. https://doi.org/10.1016/j.buildenv.2017.11.040.

[51] SANCHEZ G M E, VAN RENTERGHEM T, SUN K, et al. Using Virtual Reality for assessing the role of noise in the audio-visual design of an urban public space [J/OL]. Landscape and Urban Planning, 2017, 167: 98-107. https://doi.org/10.1016/j.landurbplan.2017.05.018.

[52] YU L, KANG J. Modeling subjective evaluation of soundscape quality in urban open spaces: an artificial neural network approach [J/OL]. The Journal of the Acoustical Society of America, 2009, 126 (3): 1163-1174. https://doi.org/10.1121/1.3183377.

[53] GEORGE D, MALLERY P. IBM SPSS statistics 23 step by step: a simple guide and reference [M]. Routledge, 2016.

[54] HANSEN J. Using SPSS for windows and macintosh: analyzing and understanding data [R/OL]. 2005. https://doi.org/10.1198/tas.2005.s139.

[55] REINTEN J, BRAAT-EGGEN P E, HORNIKX M, et al. The indoor sound environment and human task performance: a literature review on the role of room acoustics [J/OL]. Building and Environment, 2017, 123: 315-332. https://doi.org/10.1016/j.buildenv.2017.07.005.

[56] JOHN J, THAMPURAN A L, PREMLET B. Objective and subjective evaluation of acoustic comfort in classrooms: a comparative investigation of vernacular and modern school classroom in Kerala [J/OL]. Applied Acoustics, 2016, 104: 33-41. https://doi.org/10.1016/j.apacoust.2015.09.017.

[57] MACKENZIE D J, GALBRUN L. Noise levels and noise sources in acute care hospital wards

[J/OL]. Building Services Engineering Research and Technology, 2007, 28 (2): 117-131. https://doi.org/10.1177/0143624406074468.

[58] YI F, KANG J. Effect of background and foreground music on satisfaction, behavior, and emotional responses in public spaces of shopping malls [J/OL]. Applied Acoustics, 2019, 145: 408-419. https://doi.org/10.1016/j.apacoust.2018.10.029.

[59] KANG J. Acoustics in long enclosures with multiple sources [J]. The Journal of the Acoustical Society of America, 1996, 2: 985-989.

[60] KANG J. A method for predicting acoustic indices in long enclosures [J]. Applied Acoustics, 1997, 51 (2): 169-180.

[61] KANG J. The unsuitability of the classic room acoustical theory in long enclosures. Architectural Science Review 1996; 39 (2): 89-94.

[62] KANG J. Sound attenuation in long enclosures [J]. Building and Environment, 1996, 31 (3): 245-253.

3.2 The complexity of sound environment contributing to acoustic comfort in urban intermodal transit spaces

3.2.1 Introduction

Types of sounds have effects on acoustic comfort and sound enviornment in different sizes and types of spaces. Railway stations have been traditionally associated with waiting and transit spaces. In the past, this was due to the fact that they used to host a relatively limited number of functions (e. g. communicating, broadcast announcement and walking) that did not include any particular sound source. Nowadays traveling function has shifted towards new means of transport. New religion become shopping and usage of all commercial services, offered in ever-expanding terminals. It seems that architecture of transportation is now balancing between commercial and cultural function which is called urban intermodal transit spcaces[1]. In urban intermodal transit spaces, the complex acoustic environment leads to various influence among users' comfort[2]. At the same time, the preference and loudness of different sound sources are also extremely important to users. Therefore, a thorough analysis of the function of sound sources on acoustic perception and evaluation of acoustic comfort is very important to architectural researches.

Current acoustic design tend to use acoustic materials to absorb, diffuse and resonate sounds in spaces, by reaching a certain sound pressure level and reverberation time[3]. Bandyopadhyay et al. measured SPLs in the platforms and found SPLs endanger the healthful living of the users[4]. Liu et al. used acoustics testing and simulation to study the reverberation time and speech transmission index of public broadcasting system[5]. But reduction of 'sound level', does not always deliver the required improvements in quality of life[6]. In contrast, soundscape research involves human and social sciences and physical measurements for the diversity of sound environment. Moreover, it treats environmental sounds as a resource rather than a waste[7]. In recent years, soundscape was a well established approach to increase the sound quality[8,9] and also a recognised key method to managing sound environments in urban spaces[10,11]. However, there are relatively few studies investigating the quality of acoustic environments of indoor public spaces through the soundscape approach which could make important contributions to design. The absence of considerations of human perceptions and human activities from the acoustic design strategy for the transit spaces makes it a good example through which to explore the difference between traditional approaches to acoustic design and a soundscape approach to design for acoustic comfort in urban intermodal transit spaces.

This study, therefore, explores the quality of the acoustic environment in a typical urban intermodal transit station in Vienna from a soundscape perspective and discusses design strategies for achieving acoustic comfort in multi-function, open-plan intermodal transit spaces. The overall comfort level and sound environment in different functional zones were studied using a questionnaire survey. First, overall sound environment and appropriateness were analyzed. Then, the effect of different types of sound sources on acoustic comfort were analyzed. This is followed by an examination of sonic compostion effect on sound enironment and aoucstic perception in different zones.

3.2.2 Methods

3.2.2.1 Survey site

Taking a case study method[12], this study is able to explore the complexity of real life situations contributing to soundscapes and people's perceptions in a railway station. Vienna Main Station (Wien Hauptbahnhof) was selected as a typical case of urban intermodal transit spaces with mixed spatial forms, linking different transport modes and surrounding urban spaces in this area such as castles, public square, commercial area, residential area, historic area, lanes and roads. The mixed functions inside station are representative and commonly found in urban transit spaces, such as café, bars, restaurants, shops, money changing center, information board, chemist, toilet/shower, etc. Meanwhile, it has become a major urban development in its own right to include various office, retail and educational facilities. Vienna Main Station attracts a wide range of passengers and citizens, this indicates that the building is likely to have a complex acoustic environment.

Vienna Main Station has three floors, the grounded floor is the platform floor, which has 16 tracks and 15 platforms, including 5 roofed platforms and 10 platform edges. A 20,000 m² shopping centre accommodated around 100 shops and restaurants is positioned below track level, and the underground car park has spaces up to 600 cars and 1,110 bicycles. To facilitate a high rate of pedestrian movement across the station, a total of 29 escalators and 14 elevators are present to provide full step-free access to all areas[13]. In total, 800 seats spread throughout the station. A special "Kids Corner" facility is for families with young passengers, as shown in Fig. 3.9.

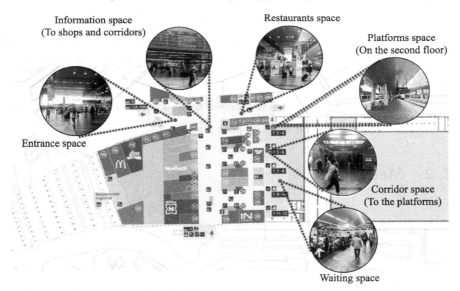

Fig. 3.9 Examples of the architectural characteristics of Vienna Main Station

3.2.2.2 Acoustic comfort survey

To study the influence of the complexity sound environment on acoustic comfort, some questionnaire survey was conducted at this case site. Every questionnaire survey was generally done by the interviewer in 3-5 min[14]. In terms of subjective investigation, 180 valid questionnaires were obtained at the survey site. To ensure the representativeness of the spaces, six typical spaces in the railway transport hub including the platform (PL), waiting space (WA), passageway (PW), entrance space (ET), restaurants space (RS) and information space (IF) were selected. The contents of the investigation concerned sound sources type, comfort evaluation and interviewees' social backgrounds[15] (Table 3.7). The results obtained from the six spaces were typical and obviously diverse.

In terms of evaluation of acoustic comfort, a five-point scale[16] was used in the questionnaire design. The evaluation of acoustic comfort was divided into five levels: 1, very uncomfortable; 2, uncomfortable; 3, neither comfortable nor uncomfortable; 4, comfortable; and 5, very comfortable. After the survey, the results of the subjective evaluation were analyzed with the software SPSS 15.0.

Table 3.7 Questionnaire questions and scales

Questions	Scale
Sex	1, male; 2, female
Age	1, <20; 2, 20-40; 3, 41-60; 4, >60
Education level	1, primary; 2, secondary; 3, higher education
Income	1, <1,000; 2, 1,000-2,000; 3, 2,001-3,000; 4, 3,001-4,000; 5, 4,001-5,000; 6, >5,000
Visit time	1, morning (9:00-11:59); 2, midday (12:00-14:59); 3, afternoon (15:00-17:59); 4, evening (18:00-21:00)
Visit duration	1, less than an hour; 2, 1-2 h; 3, more than 2 h
Evaluation of the overall sound environment	Five-point scale, with 1 being very noisy and 5 being very quiet
Evaluation of the overall acoustic comfort	Five-point scale, with 1 being very uncomfortable and 5 being very comfortable
First heard three sound sources	1; 2; 3; 4; 5
Acoustic comfort of various sound sources	Five-point scale, with 1 being very uncomfortable and 5 being very comfortable

Continued

Questions	Scale
Loudness of various sound sources	Five-point scale, with 1 being very low and 5 being very high
Intelligibility of various sound sources	Five-point scale, with 1 being very clear and 5 being very unclear
Sound level of various sound sources	Five-point scale, with 1 being very noisy and 5 being very quiet
Preference degree of various sound sources	Five-point scale, with 1 being disliked a lot and 5 being liked a lot

3.2.3 Results

On the basis of the survey results, this section discusses the influence of sound sources on sound environment and acoustic comfort. The reliability coefficient of the questionnaire was estimated as 0.83 (Cronbach's alpha). The KMO values of the subscales were greater than 0.5, and for the Bartlett spherical test, $p < 0.01$, with a reliability coefficient $0.9 > \alpha \geqslant 0.8$, the questionnaire meets the reliability[17].

3.2.3.1 Overall comfort level and sound environment

Fig. 3.10 shows the subjective evaluations of the overall sound environment and acoustic comfort in six spaces. The sound environment and acoustic comfort of the transportation hub was acceptable with the mean values of 3.81 and 3.91. It can be seen that the evaluations of sound environment and acoustic comfort in PL and ET were relatively higher (mean values of 4.23/4.42 and 4.08/4.1, respectively), and the evaluation of comfort in WA and RS were slightly lower (mean value of 3.58/3.69 and 3.36/3.48, respectively). Pearson correlation analysis between the sound environment and acoustic comfort was conducted, and the correlation coefficient was 0.683 ($p < 0.01$). This reflected that there is a significant positive correlation between sound environment and acoustic comfort. In other words, the evaluation on sound environment affected the evaluation on acoustic comfort.

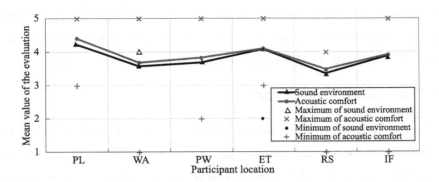

Fig. 3.10 Evaluations of the overall sound environment and acoustic comfort

There is no significant difference ($p < 0.1$) between males and females. But the age difference was significant ($p < 0.01$ or $p < 0.05$). Acoustic comfort was higher for older passengers. Education level and income difference were also significant factors ($p < 0.01$ or $p < 0.05$) for the passengers' acoustic comfort. Participants with high education level and income tend to gave positive evaluation. Differences in the frequency of visits caused a significant difference in the comfort evaluation of the sound environment in six spaces ($p < 0.05$); passengers who visited the station frequently (the mean value was 3.28) gave a more critical evaluation than passengers who did not (the mean value was 3.66). It was also found that the visit duration in a space had a significant negative correlation ($p < 0.01$) with acoustic comfort.

It is interesting to note that the evaluations on long staying spaces such as WA and RS were lower than other traffic spaces. Participants mainly reported "neither noisy nor quiet" (31.2%) and "quiet" (36.7%), however, 19.3% participants thought that the acoustic comfort was "noisy" and "very noisy". On the evaluation of acoustic comfort, they mainly reported "neither comfortable nor uncomfortable" (29.4%) and "comfortable" (38.9%), however, 16.8% participants thought that the acoustic comfort was "uncomfortable" and "very uncomfortable". The results shows that the noisier the environments is, the more uncomfortable the acoustic comfort is. This was consistent with the result of Chen and Kang on acoustic comfort on dining spaces[18], as a key factor, the background noise affected the acoustic comfort. Existing research indicated that the background noise in transport hub was an important objective index affecting passengers' acoustic comfort evaluation in the presence of composite sound sources[2]. Therefore, the following part focuses on studying the influence of sound sources in background noise on acoustic comfort evaluation.

3.2.3.2 Sonic composition

In order to identify various independent sound sources in background noise and determine the types of sound sources, participants were required to list five sound sources that they heard at that moment. Finally, various individual sound sources in six survey spaces are shown in Table 3.8.

Table 3.8 Types of sound sources in different survey spaces

Types of sound sources		Survey space					
		PL	WA	PW	ET	RS	IF
Broadcast		•			•		•
Speech sound	Speech sound of companions	•	•	•	•	•	•
	Chatting sound of other passengers	•	•	•	•	•	•
	Speech sound of staff			•		•	•
	Shouting			•		•	•
	Phone call	•	•	•	•	•	•
	Crying	•	•	•		•	
Activity sound	Footsteps		•	•	•	•	•
	Dragging luggage	•	•	•	•	•	•
	Food preparation by staff					•	
Traffic noise	Train noise	•					
Mechanical noise	Air-conditioning		•	•			
	Ventilators		•	•			
	Elevators		•	•			

Sound sources that were mentioned could be divided into five types: broadcast, speech sound, activity sound, traffic noise and mechanical noise. Speech sound and activity sound were fundamental in most spaces as key sounds. Broadcast is a common and essential sound source in the transit spaces; it can only be heard in PL, ET and IF spaces; it is easily to be covered by other sound sources in noisy spaces. Speech sound consisted of the sounds of chatting (companions, other passengers and staff) and special speech sound which were hardly merged by noises such as shouting, phone call and crying. Speech sound was mentioned as a key sound, speech sound of companions, chatting sound of other passengers and phone call could be heard in every spaces. Activity sound is the sound of users performing various activities, including footsteps, dragging luggage and food preparation by staff. Dragging luggage is a key sound that could be heard in every space, food preparation sound is a special sound source that only occurred in RS. Traffic noise is the noise made by trains as they enter and exit the station, since the platform space is on the second floor, the train noise was not easy to hear, it only appeared in PL. Mechanical noise is created during the equipment operating such as air-conditioning, ventilators and elevators.

It is interesting to note that the evaluation on sound environment and acoustic comfort in the spaces with large number of sound sources (WA, PW and RS) were lower than the spaces with relatively simple sonic composition (PL, ET and IF). The spaces that could heard various speech sound and mechanical noise made the participants felt uncomfortable and noisy. However, RS was neither had the most complex sound sources nor could heard the mechanical noise, the evaluation score on sound environment and acoustic comfort was the lowest. The reason caused the result may be the proportion of the activity sound, it can be seen from Table 3.8, three types of activity sound and all types of speech sound could be heard in RS which made the participants in the space felt more noisy and uncomfortable.

3.2.3.3 Acoustic comfort of different types of sound sources

Fig. 3.11 shows the evaluation on acoustic comfort factors in different spaces. As a key sound, the broadcast was fundamental in most spaces. The intelligibility of this sound did not get satisfied by participants both in noisy and quiet spaces, but the preference degree tended to be a comfortable score. Most participants considered the speech sound of companions as comfortable and very comfortable (38.2% and 19.8%), with the increasing of background noise, the loudness and sound level increased, the intelligibility and the preference degree decreased. The chatting sound

Indoor Sound Environment and Acoustic Perception

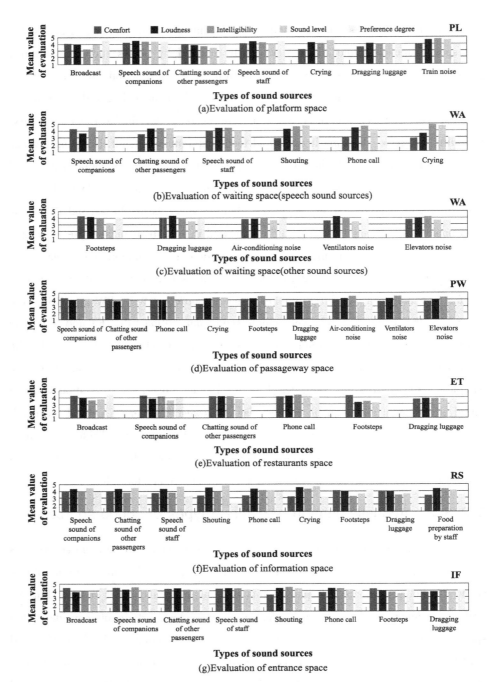

Fig. 3.11 The evaluation on acoustic comfort factors of each sound source in different spaces

of other passengers and the speech sound of staff were the same pattern. The comfort and preference degree of shouting, phone call and crying got relatively lower score. Most participants considered the shouting and crying as neither comfortable nor uncomfortable and not comfortable (33.6% and 26.7%), the reason is these two sound sources had high loudness and sound level, which may cause a decrease on the evaluation of the total acoustic environment. Although the sound level of footsteps was in a lower scale, but the loudness was high. This result in higher score on comfort and preference degree. The sound of dragging luggage is a special sound source that appears everywhere in the transit hub. The acoustic indexes of this sound source were similar in different functional spaces. Train noise was only being proposed in PL, although the loudness and sound level were high, but the comfort and preference degree were satisfied with 33.6%/28.8% felt comfortable/like and 19.2%/20.1% very comfortable and like a lot. The sound of food preparation by staff only appeared in RS, the loudness, intelligibility and sound level were very high, which caused dissatisfied on comfort and dislike on preference degree, this sound source may be the reason that cause bad evaluation on overall sound environment and acoustic comfort in RS. The evaluation of mechanical noise was relatively lower. In WA and PA, three types of mechanical noise could be heard very clear, but it is interesting to note that the comfort and preference degree of these sound sources were at a high level. In other words, mechanical noise was not the reason that cause bad evaluation on overall sound environment and acoustic comfort in these two spaces.

Table 3.9 provides a statistical analysis using the Pearson correlation between the acoustic comfort evaluation of various individual sound sources and overall sound environment in each space ($p < 0.01$). The results showed that there was a positive correlation among the acoustic comfort evaluation of shouting, phone call, crying, food preparation, train noise and overall acoustic comfort. The correlation coefficient was 0.25-0.5.

Table 3.9 Analysis of the Pearson correlation between the acoustic comfort evaluation of sound sources and the overall sound environment

Types of sound sources			Survey space					
			PL	WA	PW	ET	RS	IF
Broadcast			0.175/0.015	/	/	0.128/0.166	/	0.239/0.032
Speech sound		Companions	0.202/0.046	0.196/0.026	0.227/0.043	0.259/0.048	0.196/0.044	0.146/0.198
		Other passengers	0.317/0.062	0.258/0.035	0.168/0.066	0.283/0.136	0.183/0.117	0.219/0.043
		Staff	/	0.239/0.048	/	/	0.175/0.013	0.308/0.124
		Shouting	/	0.252/0.000 (**)	/	/	0.202/0.078	0.231/0.000 (**)
		Phone call	0.302/0.000 (**)	0.239/0.037	0.206/0.028	0.259/0.000 (**)	0.327/0.046	0.209/0.053
		Crying	0.287/0.000 (**)	0.268/0.000 (**)	0.283/0.000 (**)	/	0.215/0.209	/
Activity sound		Footsteps	/	0.302/0.082	0.208/0.036	0.248/0.027	0.196/0.023	0.236/0.092
		Dragging luggage	0.236/0.028	0.087/0.024	0.148/0.156	0.186/0.028	0.258/0.029	0.154/0.089
		Food preparation	/	/	/	/	0.369/0.000 (**)	/
Traffic noise		Train noise	0.342/0.000 (**)	/	/	/	/	/
Mechanical noise		Air-conditioning	/	0.118/0.086	0.305/0.043	/	/	/
		Ventilators	/	0.236/0.099	0.245/0.042	/	/	/
		Elevators	/	0.168/0.083	0.122/0.159	/	/	/

Note: ** representing a significance level of $p < 0.01$.

3.2.4 Conclusions

The architectural form and functions of urban intermodal transit spaces brought the complexity sound environment, lead challenges on acoustic comfort. A double height large entrance space made sound transmission between different functional spaces. Passageway spaces, waiting spaces and shops are concentrated in a low – rise and long space, it is not conducive to sound diffusion and making the acoustic environment even worse. However, the most negative influence is that the layout of different functional spaces dose not take acoustic requirements into account. In this study, on the basis of a soundscape approach by questionnaire survey conducted at an urban intermodal transit hub, the influence of various sound sources on the evaluation of sound environment and acoustic comfort were studied.

With regard to the overall sound environment and comfort level, it is found that there was a significant positive correlation between subjective comfort evaluation and sound level measurement, the more noisy the environment, the less uncomfortable the acoustic comfort. The evaluations in PL and ET were relatively higher, and the evaluation of comfort in WA and RS were slightly lower. The sonic composition of sound sources in the railway station included broadcast, speech sound, activity sound, traffic noise and mechanical noise. The dominant sound sources were not the same in each space. Broadcasts, speech sound, and activity sound were most selected by the participants. The evaluation on sound environment and acoustic comfort in the spaces with large number of sound sources (WA, PW and RS) were lower than the spaces with relatively simple sonic composition (PL, ET and IF). However, RS was neither had the most complex sound sources nor could heard the mechanical noise, the evaluation score on sound environment and acoustic comfort was the lowest.

With regard to individual sound sources, results show that the intelligibility of broadcast needed to be improved, speech sound of conversation was accepted, but shouting, phone call and crying had been paid great attention and needed to be weakened. Special sound sources that appear only in certain spaces also need attention such as train noise in PL and food preparation by staff in RS. Mechanical noise was not the reason that cause bad evaluation on overall sound environment and acoustic comfort. In order to improve acoustic environment, the sound sources of shouting, phone call, crying, food preparation, train noise were considered to have the greatest impact on the overall sound environment and acoustic comfort. Overall, this case study suggests that it is worth investigating the sound environment of urban intermodal transit spaces from a soundscape perspective might be implemented to enhance users' acoustic comfort in such spaces.

References

[1] TARCZEWSKI R, TROCKA-LESZCZYNSKA E, JABLONSKA J. Shift of the function of temple of the travelling from railway stations to airports [M] //Advances in Human Factors and Sustainable Infrastructure. Springer International Publishing, 2016.

[2] WU Y, KANG J, ZHENG W Z. Acoustic environment research of railway station in China [J]. Energy Procedia, 2018 (153): 353-358.

[3] XIAO J L, ALETTA F. A soundscape approach to exploring design strategies for acoustic comfort in modern public libraries: a case study of the Library of Birmingham [J]. Noise Mapping, 2016, 3 (1): 264-273.

[4] BANDYOPADHYAY P, BHATTACHARYA S K, KASHYAP S K. Assessment of noise environment in a major railway station in India [J]. Industrial Health, 1994, 32 (3): 187-191.

[5] LIU G, HOU D, WANG L, et al. Acoustics testing and simulation analysis of waiting hall in the line-side high-speed railway station [J]. Journal of the Acoustical Society of America 2014, 135 (4): 2332.

[6] KANG J. From dB(A) to soundscape indices: managing our sound environment [J]. Frontiers of Engineering Management, 2017, 4 (2): 184-192.

[7] KANG J, SCHULTE-FORTKAMP B. Soundscape and the built environment. Human hearing-related measurement and analysis of acoustic environments: requisite for soundscape investigations. CRC Press, 2015: 133-160.

[8] MA H, ZHANG S. A case study of soundscape design based on acoustical investigation [J]. Journal of the Acoustical Society of America, 2012, 131 (4): 3438.

[9] YU L, KANG J. Modeling subjective evaluation of soundscape quality in urban open spaces: an artificial neural network approach [J]. Journal of the Acoustical Society of America, 2009, 126 (3): 1163.

[10] MENG Q, KANG J. Effect of sound-related activities on human behaviours and acoustic comfort in urban open spaces [J]. Science of the Total Environment, 2016 (573): 481-493.

[11] MENG Q, KANG J. The influence of crowd density on the sound environment of commercial pedestrian streets [J]. Science of the Total Environment, 2015 (511): 249-258.

[12] COUSIN G. Case study research [J]. Journal of Geography in Higher Education, 2005, 29 (3): 421-427.

[13] Wien Hauptbahnh of officially inaugurated. Railway Gazette. com, 10 October (2014).

[14] LITWIN M S. How to measure survey reliability and validity [J]. The Survey Kit, 1995 (7).

[15] YU L, KANG J. Modeling subjective evaluation of soundscape quality in urban open spaces: an artificial neural network approach [J]. Journal of the Acoustical Society of America, 2009, 126 (3): 1163-1174.

[16] LIKERT R. A technique for the measurement of attitudes [J]. Archieves of Psychology, 1932, 22 (140): 1-55.

[17] GEORGE D, MALLERY P. IBM SPSS statistics 21 step by step: a simple guide and reference [M]. thirteenth ed.. Pearson, Boston, 2013.

[18] CHEN X, KANG J. Acoustic comfort in large dining spaces [J]. Applied Acoustics, 2017 (115): 166-172.

Chapter 4

Sound Environment and Acoustic Perception in Hospitals

4.1 Interaction between sound and thermal influences on patient comfort in the hospitals of China's northern heating region

4.1.1 Introduction

In complex and diverse healthcare environments, environmental comfort is a key factor that affects the evaluation of patient satisfaction[1]. The acoustic and thermal environment has the most direct influence on the patient's comfort level and satisfaction. Currently, there are a considerable number of thermal comfort studies on hospitals and other healthcare buildings. Some studies are focused on environmental parameters, such as the indoor temperature, humidity and air movement[2], while some other investigations have been presented in terms of the thermal discomfort and thermal sensation of patients and hospital staff[3-5]. According to the international standards of hospitals, the desirable indoor air temperature of regular wards is 20-24℃, and the recommended levels of relative humidity are from 30% to 60% (ISO 7330[6]). Previous research has shown that a high temperature may cause an increased out-gassing of toxins from building materials and provide more favorable growing conditions for bacteria, and low humidity can increase susceptibility to respiratory disease and contribute to irritation[7]. Therefore, the standards for temperature and humidity ranges in healthcare buildings are influenced by the measure of infection control. However, whether the thermal environment within this range is comfortable for patients seems to be ignored.

Acoustic comfort in hospital buildings has also been widely studied by researchers. Noise has been identified as a major stressor in hospitals[8]. Qin et al.[9] proposed that the general noise from nearby people was the most annoying sound, and patients' social characteristics would not significantly affect their acoustic environment evaluation. Different types of noise can be considered as acoustic pollution, and patients exposed to high amounts of noise are at a much higher risk[10]. Researchers have also measured the noise levels or studied the sound field of various healthcare environments[11,12]. Xie and Kang found that the hospital acoustic environment differed significantly every night; meanwhile, more intrusive noises tended to originate from multi-bed wards, while more extreme sounds were likely to occur in single wards[13]. Studies have found that long-term exposure to noise can lead to a range of health problems, such as an increase of cardiovascular morbidity[14] and blood pressure[15], sleep disturbances[16], and overproduction of cortisol[17]. However, the sound is not always annoying[18]. Thomas et al. showed, through a survey of a nursing home, that acoustic interventions had direct positive outcomes, as well as both positive and negative outcomes in terms of the perceived indirect effects[19].

Patients' feeling of comfort is influenced by a variety of environmental factors, and some researchers have conducted environmental interaction research. Humans perceive comfort through the interaction of various sensory stimuli and their integration in various environments. Thermal and acoustic influences are the two aspects that have the greatest influence on environmental comfort[20]. Previous research has shown that thermal comfort and acoustic comfort may influence each other[21]. A research on schools, as a case study, found that acoustic comfort was affected by thermal factors, especially in winter (Mumovic et al., 2009)[22]. Pellerin and Candas (2004) proposed that acoustic comfort and perception were affected by temperature[23]. Yang and Moon (2019) found that the impact of acoustics on indoor environmental comfort was the greatest and acoustic comfort increases under thermo neutrality[24]. However, little attention has been paid to the interactions between the sound and thermal influences inside hospitals. Meanwhile, all of these case studies were undertaken in just a single hospital, and general conclusions based on the results from just one case study are considered to be insufficient.

Therefore, the aim of the paper is to discern the interaction between the thermal and acoustic influences on environmental comfort in China's northern heating region. The general wards of 18 hospitals were investigated, and measurements of environmental factors were performed, including thermal and acoustic parameters. Thermal and acoustic evaluations were also carried out through survey questionnaires and purposely elaborated. Finally, the interaction between thermal and acoustic factors was discussed.

4.1.2 Methodology

4.1.2.1 Measurements

In China's heating region, the annual average daily temperature is at a stable ≤ 5℃ for over 90 days. The area of the heating region constitutes 70% of China's land area. The construction area of this region is about 6.5 billion m²[25]. In this paper, our investigation focuses mainly on the northern heating region of China, including the three northeastern provinces and the eastern Inner Mongolia Autonomous Region, which accounts for 51.6% of the heating region. As case studies, 18 hospitals were chosen from six cities in China's northern heating region, including Harbin, Wuchang, Qitaihe, Chifeng, Changchun and Meihekou. The measured parameters and instruments are shown in Table 4.1.

Table 4.1 Measured indoor and outdoor thermal comfort parameters, operative range and accuracy

Parameters	Definition	Measurement System	Instrument
T (℃)	Forced airy dry-bulb air temperature	ISO 7726	Testo 435
RH (%)	Relative humidity	ISO 7726	Testo 435
LAeq (dB)	Equivalent A-weighted sound pressure level	ISO 3382	BSWA 801

As for the measurement of the indoor environment, the measurement time for the current study was 8:00-18:00. Thus, it can be guaranteed that the participants will not be disturbed by activities, such as ward rounds, family visits, dining, etc., which can lead to crowd movement and affect the physical environment. These instant measurements included the temperature, relative humidity, and sound pressure level in particular hospital wards to record the environmental data while the participant responded to the survey questionnaire.

The thermal environment of the hospitals was measured from 15 November 2017 to 28 February 2018. This period provides heating for the winter in these cities. Due to the low outdoor temperature and lack of natural ventilation during heating, the airflow rate in the room is very low, so the effect of the airspeed was not considered. The air temperature and relative humidity were measured in the wards in the hospitals with Testo 435 thermo-recorders. The measuring devices were placed on the desk (0.6m above the floor) to record the thermal environment experienced by the respondent at that time. All measurement intervals were 10min.

To measure the sound environment, the equivalent continuous A-weighted sound pressure level (LAeq) was immediately recorded using BSWA 801 sound level meters. During the measurement, the sound level meters were set in slow-mode and A-weight, and reading was acquired every 3-5s. A total of 5min of data were obtained at each survey position[26]. Additionally, the distance between the measurement location, walls and other major reflective surfaces was ensured to be at least 1m, and the distance between the measurement location and the ground was 1.2-1.5m[27]. To avoid sound source variability, each sound pressure level at each measurement point was tested 10 times; each measuring point was tested every hour, and the average value of the 10 sets of data was taken as the result of this measurement point. The measuring period lasted from 8:00 to 18:00. The equipment selection and measurement process followed the ISO 3382 standard. A mean value was calculated to obtain the spatially averaged LAeq value[28].

4.1.2.2 Questionnaire survey

The participants were patients at large general hospitals in China's northern heating region. The project only deals with the evaluation of the hospital's building environment. It does not involve the investigation of the patient's condition and does not increase the pain and psychological impact of the patient's treatment process. This study does not belong to biomedical research. An ethics review and approval for this study was not required according to China's ethics, as stated in the "Methods for ethical review of biomedical research involving human beings"[29], from the Harbin

Institute of Technology's guidelines and national regulations. The research involved in the study and the plan fully considered the principles of safety and fairness. After the professor's associations from the school of architecture of the Harbin Institute of Technology reviewed the study, the rights and interests of the respondents were fully protected and the research content was guaranteed not to cause harm and risk to the respondents. The recruitment of respondents was based on the principles of voluntary and informed consent, and the subjects' rights and privacy were protected. All participants gave written informed consent. There was no conflict of interest between the research content and the research results. The survey process and questionnaire were conducted by the author (YW), a postdoctoral student with many years of experience conducting research interviews. Finally, the questionnaire was agreed upon by the expert panel, as shown in Table 4.2. The interview and research work of the project was approved to be conducted as planned. The sample comprised 220 participants (110 males and 110 females), with an average age of 49 ($SD = 15.01$).

Table 4.2 Questionnaire questions and scale

Questions	Scale
Gender	1, male; 2, female
Age	1, <20; 2, 20-40; 3, 41-60; 4, >60
The duration of stay	1, <2 days; 2, 2-5 days; 3, 6-10 days; 4, 11-20 days; 5, >20 days
The number of beds	1; 2; 4; 6; 8
Evaluation of the overall environmental comfort	Five-point scale: 1, very uncomfortable -5, very comfortable
Do you agree that the thermal environment is satisfactory?	Five-point scale: 1, strongly disagree -5, strongly agree
Do you agree that the temperature is satisfactory?	Five-point scale: 1, strongly disagree -5, strongly agree
Do you agree that the humidity is satisfactory?	Five-point scale: 1, strongly disagree -5, strongly agree
Acoustic comfort of various sound sources	Five-point scale: 1, very uncomfortable -5, very comfortable
Do you agree that the acoustic environment is satisfactory?	Five-point scale: 1, strongly disagree -5, strongly agree

Participants were interviewed individually and briefed on the purpose of the study, and they gave written informed consent to participate in the research. The interview was usually conducted in the hospital ward, but in a few cases, they were conducted in a separate room for the sake of patient privacy. Ten trained research assistants, recruited from doctor and master candidates, tested the patients. During the interviews, the participants were asked to describe situations, which they thought would bring them satisfaction or dissatisfaction in terms of the thermal and acoustic environment in which they would be treated. Field notes were written after each interview to document the immediate responses to the interaction that had occurred and aid reflexivity.

4.1.2.3 Statistical analyses

The data that we obtained in our experiment were analyzed with SPSS 15.0[30]. The Spearman correlation was used to determine the factors affecting patients' comfort evaluation of the overall environment, and the mean differences (t-test, two-tailed) in the influence of these factors on patients' comfort. The Spearman correlation analysis and regression analysis were used to determine the different building environment factors affecting patient comfort. One-way ANOVA was adopted to ascertain the factors affecting patients' comfort evaluation, including demographic and social factors, as well as the interaction between thermal and acoustic influences.

4.1.3 Results

4.1.3.1 Comfort evaluation

1. Overall comfort evaluation

According to the satisfaction evaluation of the comfort level, it should be noted that the overall comfort of the hospitals was acceptable (mean value was 4.1). The participants mainly reported "comfortable" (62.3%) and "very comfortable" (25%). However, 12.7% of the participants insisted that the satisfaction level was unacceptable (scale ≤3). Table 4.3 shows the mean and standard deviation of the patients' physical environment satisfaction evaluation. It can be seen, from column A, that the comfort of the individual physical environment factors was acceptable. The patients' satisfaction with the acoustic environment was highly acceptable (with a mean of value 4.40).

Table 4.3 Mean and standard deviation (SD) of the satisfaction with the physical environment and the overall comfort evaluation

Types of physical environment factors	A Mean and standard deviation (SD) of physical environment satisfaction	B Correlation coefficient between the factors and overall comfort
Temperature	4.04/0.911	0.686*
Relative Humidity	4.10/0.661	0.302
Acoustic	4.40/0.637	0.618*

Note: * Correlation is significant at the 0.01 level (2-tailed).

An analysis using Spearman correlation between the physical environment subjective evaluation and overall comfort was conducted, and the correlation coefficient was shown in Table 4.3 (see column B) ($p < 0.01$). The results showed that there was a significant positive correlation between the satisfaction evaluation of the temperature and acoustic with the satisfaction evaluation of the overall environmental comfort, and the temperature is more relevant than acoustics. The correlation between humidity and overall environmental satisfaction is not significant. In order to discuss the specific effects of these factors on patient comfort, the following part focuses on the influence of various individual physical environmental factors on patients' comfort evaluation.

2. Thermal evaluation

The mean indoor air temperature of the wards during the heating period was approximately 28 ℃; the temperature was kept between 25 ℃ and 31 ℃ in the wards; and the relative humidity was kept between 20% relative humidity (RH) and 50%

RH. Taking the measurement results and subjective evaluation of the temperature and RH, as shown in Fig. 4.1, it should be noted that the temperature in the wards was satisfactory, with a measured value of 26-30℃, and highly acceptable at 27-29℃. The patients had a preference for temperatures that are higher than the standard value in ISO 7730 and ASHRAE 55-2004[31,32]. Apart from the environmental parameters, two personal variables, i.e., activity and clothing level of the occupants are also very important[33]. Usually, in the ward, the patient is less active and lies in bed for a long time. In addition, the clothing is thinner, so their thermal sensation will be significantly higher than the standard comfort values. The relative humidity was unsatisfactory, and participants mainly reported dissatisfied (37.7%) and moderate (28.2%); 12.3% of the participants even claimed that they were very unsatisfied. With mean ambient temperatures around 28℃, the relative humidity is low, and heating exacerbates this phenomenon.

The measurement results of the dominant thermal parameters indicated that they may have different influences on thermal satisfaction and thermal perception. The relationship between the measured temperature and relative humidity with thermal satisfaction are also compared, where the linear regressions and coefficient of determination R^2 are also presented. It should be noted that, unlike in previous studies, in China's heating region, patients felt satisfied with an indoor temperature of 26-28℃ and a relative humidity of over 40%. With the increase of the measured temperature, the thermal satisfaction is first increased, and then drops rapidly after 28℃. The measured temperature and thermal satisfaction constituted a polynomial linear regression, and the coefficient of determination R^2 was 0.958. However, the thermal environment satisfaction increased as the relative humidity rose. The measured relative humidity and thermal satisfaction constituted an exponential linear regression, and the coefficient of determination R^2 was 0.97.

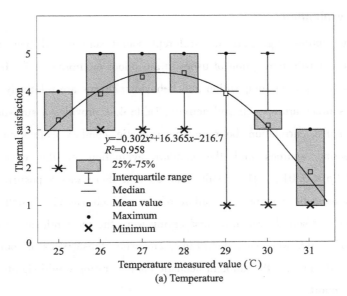
(a) Temperature

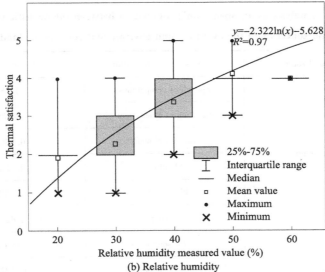
(b) Relative humidity

Fig. 4.1　The relationship between the measured temperature and relative humidity with thermal satisfaction

3. Acoustic evaluation

Previous studies suggested that different sound source and behavior patterns influence the acoustic perception of users in an indoor environment[34]. In the survey wards, the sound source composition and behavior patterns are relatively simple. The main sound sources are speech and activity. Table 4.4 provides a statistical analysis using Spearman's correlation between the acoustic comfort evaluation of various individual sound sources and the comfort evaluation of the overall acoustic environment ($p < 0.01$). The results showed that there was a positive correlation between the acoustic comfort evaluation of the speech sound of staff, shouting, crying, and pager sounds. Shouting and crying were the most relevant. In summary, these types of sounds were very disturbing, and when these types of sound were the dominant sound sources, the ambient acoustic comfort rating would significantly affect the overall comfort.

Table 4.4 Analysis using Spearman's correlation between the acoustic comfort of sound sources and the overall sound environment comfort evaluation

Types of sound sources	Sound source composition	Coefficient correlation
Speech sound	Speech of companions	0.35 *
	Chatting of other users	0.067
	Speech of staff	0.286 **
	Shouting	0.438 **
	Phone calls	0.098
	Crying	0.426 **
Activity sound	TV sounds	-0.198 *
	Pager sounds	0.392 **
	Walking sounds	0.061

Note: * Correlation is significant at the 0.05 level (2-tailed).

** Correlation is significant at the 0.01 level (2-tailed).

The mean value of LAeq in the survey wards during the daytime was 59.2dB, and LAeq was kept between 57.3dB and 63.8dB. As shown in Fig. 4.2, it should be noted that LAeq in wards was satisfactory, with the measured value of 45-65dB, and highly acceptable at 45-55dB. However, when LAeq reached 70dB, the participants mainly reported dissatisfied (43.5%) and strongly dissatisfied (21.7%). This result indicated that patients prefer a quieter environment.

Fig. 4.2 shows the relationship between the measured LAeq and acoustic satisfaction, where the linear regressions and coefficient of determination R^2 are also presented. It should be noted that LAeq lower than 65dB allowed for a satisfactory acoustic environment. With an increase in the measured LAeq, the acoustic satisfaction decreased. A correlation analysis of the evaluation of the measured value and satisfaction evaluation indicated that evaluation of the satisfaction of the acoustic environment was significantly and negatively correlated with the sound pressure level. The noisier the background noise, the lower the patients' acoustic satisfaction evaluation of the environment.

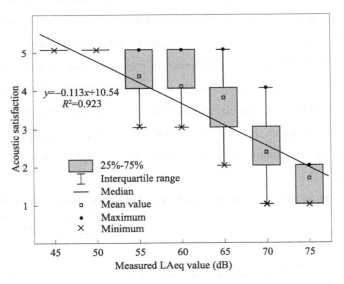

Fig. 4.2 The relationship between the measured equivalent continuous A-weighted sound pressure level (LAeq) and acoustic satisfaction

4.1.3.2 Interaction between the thermal and acoustic environment

1. The effect of the sound and thermal stimuli

Table 4.5 shows the mean rating change of the acoustic comfort evaluation, relative to the condition, with and without thermal stimuli. With the introduction of thermal factors, the mean ratings changed, compared with the condition of no thermal factors. When the temperature was low, the changes of acoustic comfort ratings were mostly less than 0.03; when the temperature was medium or high, acoustic comfort was reduced; and with the increase of temperature, the reduced decline was more obvious. No matter the type of sound source, when the humidity was low, the effect on the acoustic comfort was the most obvious, but when the humidity was medium or high, it was non-significant. In the case of shouting, crying and pager sounds, the effects of temperature and humidity were more obvious, and high temperature and low humidity had negative influences. However, with a medium temperature and humidity, the effect of improving acoustic comfort got better. In sum, the effect of thermal factors on the acoustic comfort of shouting, crying and pager sounds was the most evident, and that of speech, TV sounds and walking sounds was the least evident. Table 4.5 also shows the influence of measured time on the acoustic comfort evaluation. The occasional sound had no noticeable effect at different time periods, but the speech sounds and activity sounds show significant differences at different time periods. During the lunch break, for example, there was a significant reduction in these sounds due to human behavior, resulting in a higher score for the acoustic environment.

Table 4.5 Mean rating change of acoustic comfort, relative to the thermal condition

Indicator	Sound	Without thermal stimuli	Measured time	Thermal conditions	
				Temperature	Humidity
Speech sound	Speech sound of companions	0.18	0.15 (8-10am), 0.12 (10-12am), 0.36 (12am-2pm), 0.24 (2-4pm), 0.13 (4-6pm)	0.15 Low, 0.17 Medium, 0.13 High	0.02 Low, 0.21 Medium, 0.18 High

Chapter 4 Sound Environment and Acoustic Perception in Hospitals

Continued

Indicator	Sound	Without thermal stimuli	Measured time	Thermal conditions — Temperature	Thermal conditions — Humidity
Speech sound	Chatting sound of other users	0.11	8-10am: 0.09; 10-12am: 0.07; 12am-2pm: 0.23; 2-4pm: 0.18; 4-6pm: 0.08	Low: 0.07; Medium: 0.09; High: −0.21	Low: 0.13; Medium: 0.08; High: 0.11
	Speech sound of staff	0.22	8-10am: 0.12; 10-12am: 0.13; 12am-2pm: 0.39; 2-4pm: 0.26; 4-6pm: 0.24	Low: 0.23; Medium: 0.25; High: 0.11	Low: 0.09; Medium: 0.21; High: 0.19
	Shout	−0.23	8-10am: −0.39; 10-12am: −0.28; 12am-2pm: −0.18; 2-4pm: −0.02; 4-6pm: −0.09	Low: −0.21; Medium: −0.25; High: −0.39	Low: −0.28; Medium: −0.19; High: −0.17
	Phone call	−0.11	8-10am: −0.12; 10-12am: −0.19; 12am-2pm: −0.05; 2-4pm: −0.08; 4-6pm: −0.17	Low: −0.08; Medium: −0.09; High: −0.17	Low: −0.13; Medium: −0.07; High: −0.05
	Cry	−0.34	8-10am: −0.35; 10-12am: −0.38; 12am-2pm: −0.29; 2-4pm: −0.29; 4-6pm: −0.32	Low: −0.28; Medium: −0.31; High: −0.45	Low: −0.38; Medium: −0.32; High: −0.25
Activity sound	TV sound	0.11	8-10am: 0.09; 10-12am: 0.13; 12am-2pm: 0.08; 2-4pm: 0.16; 4-6pm: 0.11	Low: 0.19; Medium: 0.15; High: 0.03	Low: 0.08; Medium: 0.03; High: 0.04
	Pager sound	−0.02	8-10am: 0.03; 10-12am: 0.02; 12am-2pm: −0.08; 2-4pm: 0.01; 4-6pm: −0.02	Low: 0.08; Medium: 0.16; High: −0.07	Low: −0.09; Medium: −0.03; High: −0.02
	Walking sound	0.02	8-10am: 0.06; 10-12am: 0.02; 12am-2pm: −0.09; 2-4pm: 0.05; 4-6pm: 0.04	Low: 0.03; Medium: 0.02; High: −0.04	Low: −0.04; Medium: −0.01; High: 0.04

2. The interaction between the thermal and acoustic influences on comfort

Despite the results presented in the literature[35], where thermal comfort is in general ranked by the building occupants as having a greater importance than visual and acoustic ones, in this survey, the thermal and acoustic factors had about the same weight. Existing research indicated that indoor environments were formed under the combined action of various physical environmental factors[36]. The interaction of these factors should be taken into consideration.

Mixed design analyses of variance (ANOVAs) were run to test the influence of different temperatures and humidity levels on the sound pressure level. As shown in Table 4.6, the significance level (Sig.) is the p-value of the variance F test. $p \leqslant 0.05$ means that the temperature and humidity significantly influence LAeq at a level of 0.05. The deviation means square of the sound pressure level for different temperatures and humidity levels ($A \times B$) is 18.902; the F value is 1.418; and the significance level is 0.298, that is, $p > 0.05$ shows no significant difference. The results of the multiple comparison analysis under the interaction of the sound pressure level and temperature showed that there was no significant difference between 18℃ and 22℃ ($p > 0.05$), as shown in Table 4.7. The interaction between LAeq and temperature showed that with the increase of temperature, LAeq was significantly increased. The reason may be that people experienced more negative emotions when the thermal environment deviated from neutral conditions[37], the volume was automatically increased, or the ratio of shouting and crying was higher.

Table 4.6　Mixed design analyses of variance (ANOVAs)　　Dependent variable: LAeq

Source	Mean square	F	Sig.
Temperature (A)	1575.423	90.815	0.000
Humidity (B)	313.087	18.512	0.000
Temperature × Humidity ($A \times B$)	18.902	1.428	0.298

Table 4.7　Multiple comparisons analyses　　Dependent variable: LAeq

Temperature (I)	Temperature (J)	Mean difference (I-J)	St. error	Sig.	95% Confidence interval	
					Lower bound	Upper bound
18	22	6.018	1.700	0.116	-1.23	13.266
	26	18.176**	1.700	0.000	12.968	23.384
	30	26.897**	1.700	0.000	17.616	36.178

Continued

Temperature (I)	Temperature (J)	Mean difference (I-J)	St. error	Sig.	95% Confidence interval	
					Lower bound	Upper bound
22	18	-6.018	1.700	0.116	-13.266	1.23
	26	12.088*	1.700	0.000	10.665	13.511
	30	15.286*	1.700	0.000	9.683	20.889
26	18	-18.176*	1.700	0.000	-23.384	-12.968
	22	-12.088*	1.700	0.000	-13.511	-10.665
	30	9.575*	1.700	0.002	3.981	15.169
30	18	-26.897*	1.700	0.000	-36.178	-17.616
	22	-15.286*	1.700	0.000	-20.889	-9.683
	26	-9.575*	1.700	0.002	-15.169	-3.981

Note: * The mean difference is significant at the 0.01 level.

4.1.4 Discussion

The factors affecting patients' comfort in hospital wards include factors other than acoustic and thermal influences. In the following paragraphs, some major factors are discussed, which must be reviewed and updated in future studies.

House-related factors. Aspects related to the house configuration significantly affect the indoor environment. The room location (facing the street vs. facing a quiet side) was found to be correlated with road traffic noise annoyance. In this study, the room location has a significant influence on acoustic comfort and thermal comfort. Rooms facing the streets were found to negatively affect annoyance caused by the passage of nearby traffic. However, due to these rooms facing south, bringing in more light, indoor thermal conditions were rated relatively higher than those facing north.

Fig. 4.3 provides the measured LAeq over the test period with different ward capacities. Since the density of patients and the size of the space affect the indoor thermal and acoustic environment, we investigated wards with different numbers of beds. It should be noted that the environmental noise in six-bed rooms and eight-bed rooms were higher than that in single, double and quad rooms. This indicated that the patient density had a significant impact on environmental noise, which is consistent with a study of an underground commercial street[39]. The noise level of each ward starts to rise at 8:00 and becomes relatively quiet from 12:00 to 14:00 because of the lunch break. The peak noise at 70-75dB occurs at 9:00-11:00 and 15:00-17:00, and this level of noise can be annoying or uncomfortable. As in previous studies, more occurrences of noises with longer duration were observed in multiple-bed wards, as compared with single-bed wards[40]. This indicates that the measurement results of the acoustic environment were similar to those in different climate regions.

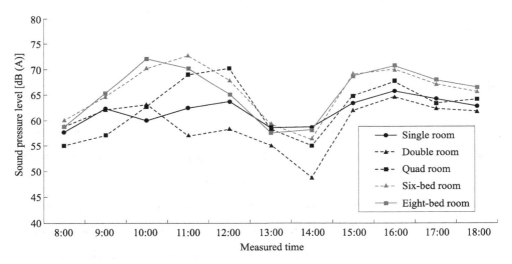

Fig. 4.3 Measured LAeq over the test period with different ward capacities

Person-related factors. Gender and age are factors that must be taken into account in the assessment of comfort in a hospital[41]. The mean difference between males and females in the evaluation of thermal and acoustic comfort was determined, and there is no significant difference ($p < 0.1$) between males and females in terms of acoustic comfort, with a similar mean value (female 4.21 and male 4.12), as shown in Fig. 4.4. These results were consistent with those of previous studies, which suggested that the effect of gender on sound annoyance evaluation is generally insignificant[42-44]. There are significant gender differences in thermal comfort, as shown in Fig. 4. 5. Females are more critical than males about temperature, and females prefer higher room temperatures than males.

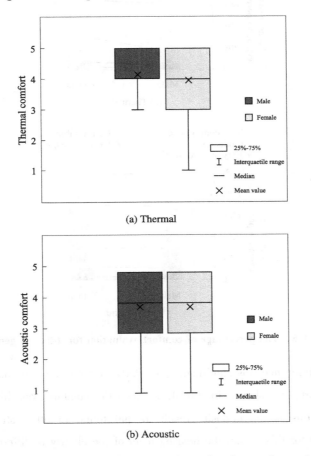

Fig. 4. 4 **The gender difference on thermal and acoustic comfort**

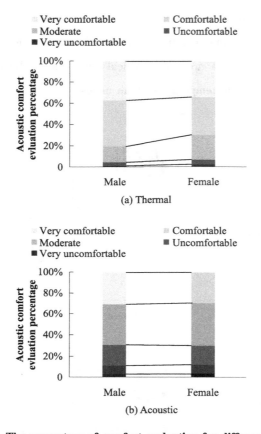

Fig. 4.5 The percentage of comfort evaluation for different genders

The age difference was significant ($p < 0.01$ or $p < 0.05$) for both thermal and acoustic comfort, as shown in Fig. 4.6. Acoustic comfort was higher for older patients, and this is the opposite result to that found in other studies[45,46]. The possible reason for this is that the hearing level of the elderly is different from that of the young. Younger people were found to be highly annoyed by noise, which are the same results as those found in previous studies[47]. A hot and dry environment was considered to be more acceptable among elders than among young patients as shown in Fig. 4.7. A previous study drew the opposite conclusion, namely, that thermal comfort was neither influenced by gender nor age[48].

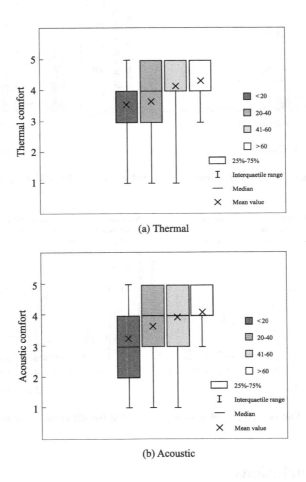

Fig. 4.6 The age difference on thermal and acoustic comfort

It is interesting to note that the duration of stay has a significant negative effect on thermal and acoustic comfort evaluation. As the stay length increased, the patients gave more negative comments, as shown in Fig. 4.8.

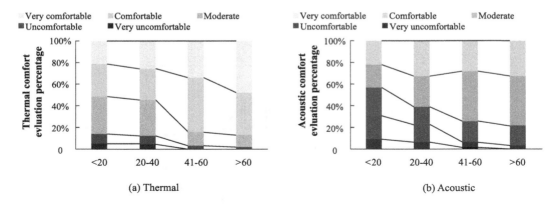

Fig. 4.7 The percentage of comfort evaluation for different ages

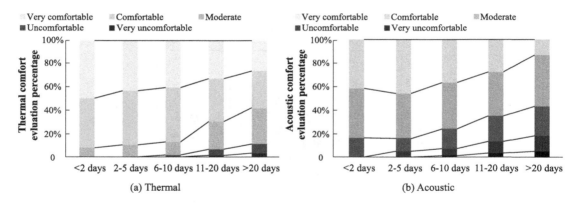

Fig. 4.8 The percentage of comfort evaluation for different durations of stay

4.1.5 Conclusions

This study investigated the interaction between sound and thermal comfort in hospital wards in China's heating region. An objective investigation was carried out, where participants evaluated sound comfort, thermal comfort and overall comfort. Meanwhile, a subjective measurement of the sound pressure level, temperature and humidity were carried out.

For overall comfort, the effects of sound and temperature are stronger than the effect of humidity, while the effects of sound and temperature are almost equal. For thermal comfort, the temperature in the wards was satisfactory, with a measured value of 26-30℃, and highly acceptable at 27-29℃. The patients had a preference for temperatures that are higher than the standard value in ISO 7730 and ASHRAE 55-2004. The relative humidity was unsatisfactory, and heating exacerbates this phenomenon. For acoustic comfort, LAeq in the wards was satisfactory, with a measured value of 45-65dB, and highly acceptable at 45-55dB. There was a positive correlation between the acoustic comfort evaluation of the speech of staff, shouting, crying and pager sounds, with shouting and crying being the most relevant.

The influence of thermal factors on sound evaluation showed that a low temperature has little effect on the evaluation of acoustic comfort, whereas for any type or volume of sounds, the higher the temperature is, the more negative the evaluations is. Irrespective of the type of sound source, when the humidity was low, the effect on acoustic comfort was the most obvious, but when the humidity was medium or high, it was non-significant. In terms of shouting, crying and pager sounds, the effects of temperature and humidity were more obvious, and high temperature and low humidity have a negative influence. With a medium temperature and humidity, the effect of improving acoustic comfort got better.

The similarities and differences in the evaluations of the sound and thermal influences also showed that there is an analogous trend in patients' environment comfort and preferences because of their high correlation coefficient. A satisfactory thermal environment can improve the evaluation of acoustic comfort, while an unsatisfactory thermal environment has the opposite effect.

Notwithstanding its limitation, this study clearly indicates that the special climatic environment in cold regions of China can cause changes in patient satisfaction. The building environments that affect patients' satisfaction include the thermal and acoustic environment. It should be noted that this study was focused specifically on patients in the cold region of China, and this specificity could affect the generalization of their findings of other patients. The geographical limitations of this study should be considered in interpreting these findings. Other regions or countries may produce different results. The three influencing factors selected in this study are dominated by thermal and acoustic comfort. Because of the dominant role of vision and air quality, future research can add lighting factors and indoor air quality as variables to study the influence of these additional factors on patient satisfaction.

References

[1] SILVA P L, PAIVAL, FARIA V B, et al. Triage in an adult emergency service: Patient satisfaction [J]. Revista da Escola de Enfermagem da Usp, 2016, 50 (3): 427-432.

[2] MURPHY J. Temperature and humidity control in surgery rooms. Ashrae Journal, 2006, 48 (6): 18-25.

[3] YAU Y H, CHEW B T. Thermal comfort study of hospital workers inMalaysia [J]. Indoor Air, 2009, 19 (6): 500-510.

[4] ASHRAE (Firm). Standard 55-2013 user's manual: ANSI/ASHRAE standard 55-2013. In Thermal Environmental Conditions for Human Occupancy [S]. ASHRAE Research: Atlanta, GA, USA, 2016.

[5] POURSHAGHAGHY A, OMIDVARI M. Examination of thermal comfort in a hospital using PMV-PPDmodel [J]. Applied Ergonomics, 2012, 43 (6): 1089-1095.

[6] ISO. Moderate Thermal Environments—Determination of the PMV and PPD Indices and Specifications for Thermal Comfort [S]. 2nd ed. International Organization for Standardization: Geneva, Switzerland, 2005.

[7] KAMEEL R, KHALIL E. Thermal comfort VS air quality in air-conditioned healthcare applications [M]. In Proceedings of the 36th AIAA Thermophysics Conference, Orlando, FL, USA, 2003: 2003-4199.

[8] FARREHI P M, NALLAMOTHU B K, NAVVAB M. Reducing hospital noise with sound acoustic panels and diffusion: A controlled study [J]. BMJ Quality & Safety, 2015, 25 (8): 644-646.

[9] QIN X, KANG J, JIN H. Subjective evaluation of acoustic environment of waiting areas in general hospitals [J]. Building Science, 2011, 12: 53-60.

[10] OLIVERA J M, ROCHA L A, ROTGER V I, et al. Acoustic pollution in hospital environments [J]. Journal of Physics Conference Series, 2011, 332.

[11] SHIERS N J, SHIELD B M, GLANVILLE R E. Patient and staff perceptions of hospital noise [J]. The Journal of the Acoustical Society of America, 2012, 132 (3): 2032.

[12] VAN REENEN C A. A case study investigation of the indoor environmental noise in four urban South African hospitals [J]. The Journal of the Acoustical Society of America,

2015, 138 (3): 1900.

[13] XIE H, KANG J. The acoustic environment of intensive care wards based on long period nocturnal measurements [J]. Noise and Health, 2012, 14 (60): 230-236.

[14] BELOJEVIC G, PAUNOVIC K, JAKOVLJEVIC B, et al. Cardiovascular effects of environmental noise: Research in Serbia [J]. Noise and Health, 2011, 13 (52): 217-220.

[15] BABISCH W, SWART W, HOUTHUIJS D, et al. Exposure modifiers of the relationships of transportation noise with high blood pressure and noise annoyance [J]. The Journal of the Acoustical Society of America, 2012, 132 (6): 3788-3808.

[16] TERZANO M G, PARRINO L, FIORITI G, et al. Modifications of sleep structure induced by increasing levels of acoustic perturbation in normal subjects [J]. Electroencephalography and Clinical Neurophysiology, 1990, 76 (1): 29-38.

[17] RECIO A, LINARES C2, BANEGAS JR, DÍAZ J. Road traffic noise effects on cardiovascular, respiratory, and metabolic health: An integrative model of biological mechanisms [J]. Environmental Research, 2016, 146: 359-370.

[18] TORRESIN S, ALBATICI R, ALETTA F, et al. Assessment methods and factors determining positive indoor soundscapes in residential buildings: A systematic review [J]. Sustainability, 2019, 11 (19): 5290.

[19] THOMAS P, ALETTA F, FILIPAN K, et al. Noise environments in nursing homes: An overview of the literature and a case study in Flanders with quantitative and qualitative methods [J]. Applied Acoustics, 2020, 159: 107103.

[20] SILVA S M, ALMEIDA M G D. Thermal and acoustic comfort in buildings [J]. INTER-NOISE and NOISE-CON Congress and Conference Proceedings, InterNoise10, Lisbon, Portugal, 2010: 6967-6976.

[21] KARMANN C, BAUMAN F S, RAFTERY, P, et al. Cooling capacity and acoustic performance of radiant slab systems with free-hanging acoustical clouds [J]. Energy and Buildings, 2017, 138: 676-686.

[22] MUMOVIC D, PALMER J, DAVIES M, et al. Winter indoor air quality, thermal comfort and acoustic performance of newly built secondary schools in England [J]. Building and Environment, 2009, 44 (7): 1466-1477.

[23] PELLERIN N, CANDAS V. Effects of steady-state noise and temperature conditions on environmental perception and acceptability [J]. Indoor Air, 2004, 14 (2): 129-136.

[24] YANG W, MOON H J. Combined effects of acoustic, thermal, and illumination conditions on the comfort of discrete senses and overall indoor environment [J]. Building and

Environment, 2019, 148 (15): 623-633.

[25] LIANG C, LU M, WU Y. Research on indoor thermal environment in winter and retrofit requirement in existing residential buildings in China's northern heating region [J]. Energy Procedia, 2012, 16, PartB: 983-990.

[26] MENG Q, ZHAO T, KANG J. Influence of Music on the Behaviors of Crowd in Urban Open Public Spaces [J]. Frontiers in Psychology, 2018, 9: 596.

[27] BARRON M, FOULKES T J. Auditorium acoustics and architectural design [J]. The Journal of the Acoustical Society of America, 1994, 96 (1): 612.

[28] ZHANG Y, ZHAO Y W, YANG Z F, et al. Measurement and evaluation of the metabolic capacity of an urban ecosystem [J]. Communications in Nonlinear Science and Numerical Simulation, 2009, 14 (4): 1758-1765.

[29] Methods for Ethical Review of Biomedical Research Involving HumanBeings [Z/OL]. Bulletin of the State Council of the People's Republic of China. 2017 (No. 27). [2019-09-02]. http://www.gov.cn/gongbao/content/2017/content_5227817.htm.

[30] HANSEN J. Using SPSS for windows and macintosh: Analyzing and understanding data [J]. The American Statistician, 2005, 59 (1): 113.

[31] ISO. Moderate Thermal Environment—Determination of PMV and PPD Indices and Specifications of the Conditions for Thermal Comfort [S]. International Organization for Standardization: Geneva, Switzerland, 1994.

[32] ASHRAE Standard. Thermal environmental conditions for human occupancy [S]. American Society of Heating, Refrigerating and Air-Conditioning Engineers, 1992.

[33] SINGH M K, MAHAPATRA S, ATREYA S K. Adaptive thermal comfort model for different climatic zones of North-East India [J]. Applied Energy, 2011, 88 (7): 2420-2428.

[34] WU Y, KANG J. Acoustic environment research of railway station in China [J]. Energy Procedia, 2018, 153: 353-358.

[35] NEMATCHOUA M K, RICCIARDI P, BURATTI C. Statistical analysis of indoor parameters an subjective responses of building occupants in a hot region of Indian ocean: A case of Madagascar Island [J]. Applied Energy, 2017, 208 (15): 1562-1575.

[36] BURATTI C, BELLONI E, MERLI F, et al. A new index combining thermal, acoustic, and visual comfort of moderate environments in temperate climates [J]. Building and Environment, 2018, 139: 27-37.

[37] LAN L, LIAN Z. Use of neurobehavioral tests to evaluate the effects of indoor environment quality on productivity [J]. Building and Environment, 2009, 44 (11): 2208-2217.

[38] RIEDEL N, KÖCKLER H, SCHEINER J, et al. Home as a place of noise control for the

elderly? A cross-sectional study on potential mediating effects and associations between road traffic noise exposure, access to a quiet side, dwelling-related green and noise annoyance [J]. International Journal of Environmental Research & Public Health, 2018, 15 (5): 1036.

[39] MENG Q, KANG J, JIN H. Field Study on the Influence of Spatial and Environmental Characteristics on the Evaluation of Subjective Loudness and Acoustic Comfort in Underground Shopping Streets [J]. Applied Acoustics, 2013, 74 (8): 1001-1009.

[40] XIE H, KANG J, MILLS G H. Behavior observation of major noise sources in critical care wards [J]. Journal of Critical Care, 2013, 28 (6): 1109. e5-1109. e18.

[41] DEL FERRARO S, IAVICOLI S, RUSSO S, et al. A field study on thermal comfort in an Italian hospital considering differences in gender and age [J]. Applied Ergonomics, 2015, 50: 177-184.

[42] MENG Q, KANG, J. Effect of sound-related activities on human behaviours and acoustic comfort in urban open spaces [J]. Science of The Total Environment, 2016, 573: 481-493.

[43] LI H N, CHAU C K, TANG S K. Can surrounding greenery reduce noise annoyance at home? [J]. Science of The Total Environment, 2010, 408 (20): 4376-4384.

[44] AMUNDSEN A H, KLAEBOE R, AASVANG G M. The Norwegian Façade Insulation Study: The efficacy of façade insulation in reducing noise annoyance due to road traffic [J]. The Journal of the Acoustical Society of America, 2011, 129 (3), 1381-1389.

[45] CHUNG W K, CHAU C K, MASULLO M, et al. Modelling perceived oppressiveness and noise annoyance responses to window views of densely packed residential high-rise environments [J]. Building and Environment, 2019, 157: 127-138.

[46] PAIVA K M, CARDOSO M R A, ZANNIN P H T. Exposure to road traffic noise: Annoyance, perception and associated factors among Brazil's adult population [J]. Science of The Total Environment, 2019, 650 (Part 1): 978-986.

[47] JANSSEN S A, VOS H, EISSES A R, ET AL. A comparison between exposure-response relationships for wind turbine annoyance and annoyance due to other noise sources [J]. The Journal of the Acoustical Society of America, 2011, 130 (6): 3746-3753.

[48] FRONTCZAK M, WARGOCKI P. Literature survey on how different factors influence human comfort in indoor environments [J]. Building and Environment, 2011, 46 (4): 922-937.

4.2 Influence of the acoustic environment in hospital wards on patient physiological and psychological indices

4.2.1 Introduction

The indoor environment as a service carrier is most directly influenced by mental feelings, which are linked to patient comfort and mood. Patients stay in the ward almost all day, highlighting the great importance of providing comfortable conditions in hospital buildings. The comfort environment is considered the most important factor influencing patient feelings (Buckles, 1990). Researchers have measured the noise levels or studied the sound source of various health care environments, such as critical care wards (Xie et al., 2013), intensive care units (Xie et al., 2009), and entrance halls (Qin et al., 2011). As shown in previous studies, the noise levels measured in the wards frequently exceed the World Health Organization guideline values [45dB(A)] by more than 20dB(A) (Berglund et al., 1999; MacKenzie and Galbrun, 2007). Noise is a major public health issue, and noise annoyance is the most common and direct response among people exposed to environmental noise. Noise has been identified as a major stressor in hospitals (Farrehi et al., 2016), and will influence an individual's physical and mental health.

4.2.1.1 Literature review

The documented association with several diseases and the growing number of exposed persons worldwide (Recio et al., 2016) indicate negative emotional and attitudinal reactions to noise (Okokon et al., 2015). Exposure to noise may interfere with daily activities, feelings, thoughts, rest or sleep, and may be accompanied by negative emotional reactions, such as irritability, distress, exhaustion and other stress-related symptoms (Beutel et al., 2016). The impacts of stressors on health depend on the complex interactions between stressors and individual coping strategies, which are developed through previous experience, psychology, biology, social factors, competitive stressors and personality (Jensen et al., 2018). Noise-related health problems are growing, and more severe effects related to cardiovascular morbidity and mortality have been proposed (Belojevic et al., 2011). Studies have found that an increase in daily noise levels of 1dB (1) resulted in a 6.6% increase in the risk of death in the elderly (Tobías et al., 2015) and have observed a significant increase in blood pressure of 2-4 mmHg after 10 min of high-level exposure (Paunovi et al., 2014). Studies have also reported that noise is negative impacts mental health (Hammersen et al., 2016; Jensen et al., 2018), which interacts with a wide range of complex elements, including biological, psychological, social, economic and environmental factors (Barry and Friedli, 2010; Aletta et al., 2017). These factors include not only objectively measured environmental conditions but also subjective evaluation. When the noise level can no longer be reduced, people can still be annoyed by the noise because their subjective feelings can be affected by other psychoacoustic attributes, such as sharpness and roughness (Zwicker and Fastl, 1999). If noise-induced annoyance is a chronic problem (perceived as little or uncontrollable), it might cause not only stress but also fatigue associated with ineffective attempts to cope with noise and then impact mental health (Eisenmann, 2006). Noise and noise annoyance have nonstandard effects on individuals that might depend on previous experiences or biological susceptibility. When individuals do not have control over the noise, as experienced with noise annoyance, individuals might suffer from learned helplessness and biological signatures of chronic stress, including overproduction of cortisol (Recio et al., 2016).

Since the formulation of eco-effective design (EED) and evidence-based design, the restorative effects of the environment have attracted wide attention (Ulrich et al., 2008; Shepley et al., 2009). As the primary facility for helping people to recover from illness, hospitals have also begun to focus on developing a healthy spatial environment utilizing natural forces. Through studying soundscapes, sounds in the environment have been regarded as a useful resource, and a favorable and healthy spatial environment can be created through discussing human perception and experience (Kang et al., 2016). As a result of people's perception of the acoustic environment, soundscapes can be positive (such as happy, calm) or negative (such as worry, pressure). The research on the effect of soundscape restoration is based on the development of attention restoration theory (ART) proposed by Kaplan et al. (1989) and stress restoration theory (SRT) proposed by Ulrich (Ulrich et al., 1991). Weakening negative soundscapes is significantly related to health status, and increasing positive soundscapes is significantly related to environmental pressure recovery (Aletta et al., 2018a, b). A series of previous studies revealed that design and occupant choices can have positive health impacts by controlled reduction of noise levels (Evans, 2003; Von Lindern et al., 2016; Aletta et al., 2018c). It was also found that the natural environment had a positive effect on restoration processes (Hartig and Staats, 2003). However, the restorative effects of soundscapes should not only correlated with subjective evaluation data but also with physiological parameters, including the emotions caused by sound stimulation (Hume and Ahtamad, 2013; Aletta et al., 2016a). Moreover, the soundscape is related to other spatial environmental factors. When a person hears a sound, the perceived auditory space around them may modulate their emotional response to it. Small rooms are considered to be more pleasant, calmer and safer than large rooms, and sounds originating behind listeners tend to be more arousing and elicit larger physiological changes than sources in front of the listeners (Tajadura-Jiménez et al., 2010). In their work on soundscapes in hospitals, researchers have revealed the relationship between the acoustic environment, typical sound sources and geometry form (Xie and Kang, 2012a). An acoustic environment evaluation system has also been established (Xie and Kang, 2012b), and it has been found that the acoustic environment plays a leading role in the overall environmental evaluation has also been found (Wu et al., 2019). However, the impacts of hospital acoustic soundscapes on the physiological and psychological indices of patients require further study.

Perceptual experiences in one modality often depend on activity from other sensory modalities. The renewed interest in the topic of crossmodal correspondences that have emerged in recent years has motivated research that demonstrated that crossmodal matchings and mappings exist between most sensory dimensions (Deroy and Spence, 2016). Individuals reliably match different tastes/flavors (KnFerle and Spence, 2012), colors (Hamilton et al., 2016), and shapes (Ozturk et al., 2013) to auditory stimuli. For example, individuals consistently match high-pitched sounds to small, bright objects located high up in space (Spence, 2011). In each experimental module, participants were experiencing different hospital indoor environments as the different experimental scenario conditions. Experimental scenarios can be classified as real or artificial. Due to site restrictions, it is difficult to effectively control a large number of irrelevant environmental factors, and it is also difficult to "add" a new environmental factor to the original indoor environment. Therefore, the experimental conditions of real scenarios are limited by their controllability (Stamps III, 2007). To gain the most accurate and least variable estimate of acoustic environmental stimuli/properties, the stimulation of other environmental factors should be minimized. With the increasing maturity of virtual reality (VR) technology in recent years, VR environments can provide users with a more realistic and immersive environment (Chamilothori et al., 2019). Multiple empirical studies show that the physiological, psychological and behavioral feedback of participants in VR scenarios are similar to those in real scenarios (Heydarian et al., 2015). Yin et al. (2018) found that, in VR scenarios, the user's heart rate, blood pressure, skin conductivity, cognitive ability and emotional level were very similar to those in real scenarios. Therefore, environmental psychologists began to use VR scenarios for environmental psychology experiments, rather than real scenarios.

4.2.1.2 Study's framework

This study aims to determine the following: (1) whether the acoustic environment can affect recovery in terms of physiological indicators; we hypothesize that physiological recovery will increase with music and decrease with artificial sounds and mechanical sounds; (2) whether sounds can decrease or increase the psychological function of patients in hospital wards; we hypothesize that music will be helpful for the psychological restoration of patients, as artificial and mechanical sounds will lead to the opposite trend; and (3) whether demographic factors and other environmental factors will cause different degrees of impact; we hypothesize that differences in demographic and environmental factors will lead to differences in the degree of the effect of soundscape recovery, as some previous studies indicated that there are differences between population and other environmental factors in the subjective evaluation of the acoustic environment. A digital three-dimensional (3D) model of a room was constructed, and experimental patients wore VR glasses to visualize the same ward scene and eliminate other visual and landscape distractions. Several different approaches were explored to meet the goals. First, the effect of sound stimuli on the physiological indices of the patients was examined. Second, the effect of sound stimuli on an individual's mental health was examined. Third, differences in the effects of sound on different populations and multiple environmental interactions were observed.

4.2.2 Methodology

In this study, a combination of physiological measurements and psychological evaluation was utilized. Four typical sound types were presented to experimental patients and their physiological indicators were monitored by attached detectors. The patients wore VR glasses to observe the same virtual ward space and eliminate interference from other environments. The participants were asked to complete a subjective questionnaire. The obtained data were analyzed to evaluate the restorative effect of sounds in hospitals on individuals utilizing statistical methods.

4.2.2.1 Participants

The participants were all inpatients of the internal medicine department of the First Hospital of Harbin and the second and fourth affiliated hospitals of Harbin Medical University. Inpatients from internal medicine department tend to have more time to participate in experiments than outpatients, and internal medicine is mainly related to chronic diseases; this provides an ideal experimental object that can exclude the psychological and physiological effects of diseases.

The participants were 70 patients with an average age of 48.2 ($SD = 3.42$; min = 18; max = 72), including 36 men and 34 women. The determination of the participants' proportions excluded the effects of differences in participants' composition on the experimental results. (Zhang et al., 2016). The number of participants selected in this study was based on relevant experiments conducted in similar fields (Alvarsson et al., 2010; Annerstedt et al., 2013).

All participants were required to have clear cognitive consciousness and sufficient visual, auditory and behavioral abilities to ensure that they could complete the physiological index measurement and questionnaire survey, and wore comfortable clothing. Additionally, patients with hyperthyroidism and supraventricular tachycardia were excluded from this experiment as autonomic nervous dysfunction would decrease the accuracy of the measurement and evaluation of physiological stress indicators. The diet and sleep status of the participants needed to be stable. Six hours before the test, the patients did not drink, smoke, or have coffee or other drinks that would stimulate the sympathetic nervous system (Li and Kang, 2019).

The study was approved by the professors' associates in the School of Architecture at Harbin Institute of Technology. Written informed consent was obtained from all participants before the test began. Participants were informed about the goals and contents of the study, privacy and data protection, and that their participation in the study was voluntary. Biological samples were not collected.

4.2.2.2 Visual scene

Many studies have been conducted on the influence of audio-visual factors on noise perception (Collignon et al., 2008; Yost and Zhong, 2015; Aletta et al., 2016b; Liu and Kang, 2018), and it has been demonstrated that vision and hearing can influence one other. Therefore, to prevent other factors influencing the patients' psychological and physiological indices, participants wore VR glasses and observed the same ward scene. A standard single ward was selected as the experimental scene, as shown in Fig. 4.9 (a). The simulated ward volume was 6.6m × 3.6m × 2.8m, and the bed size was 2.1m × 0.9m. A U-shaped rail curtain and 0.45m × 0.45m bedside cabinet were set around the bed. Participants experienced the scene from the perspective of sitting on the bed in the ward and wore VR headsets, as shown in Fig. 4.9 (b). The transformation of the virtual environment was based on the experimental condition transformation path of basic model construction—experimental parameter adjustment—virtual scene generation. First, a digital 3D model of the indoor space was established by 3DMAX. Then, according to the specific experimental goal, some of the design parameters of the model scenario were adjusted. Finally, the adjusted model was imported into the HTC Vive VR device.

 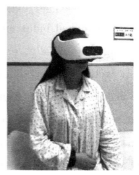

(a) Virtual scene of the ward　　　　　　(b) Participant wearing VR

Fig. 4.9　**Virtual reality device HTC Vive Focus Plus and presenting virtual scene**

4.2.2.3　Selection of sound stimuli

Common hospital sound sources can be summarized into four typical categories: mechanical, artificial, background and music (Rashid and Zimring, 2008). Mechanical sounds are produced by hospital equipment, such as wheelbarrows, ventilators and electrocardiograph monitors. Artificial sounds include patient conversations, children's crying, phones being answered and other behaviors. Background sounds have no clear dominant source, and include mechanical sounds produced by new air systems, elevator operations and other equipment, as well as artificial sounds produced by the conversations and movements of doctors and patients. Mechanical, artificial and natural sounds were recorded in the participants' hospital. According to Baker's classification standard of hospital background noise, the background should be stable, with its amplitude changing less than once per minute (< 5 dB) (Baker, 1992). "Day and Night" was selected for the music stimuli, which is a particular piece of music that has been widely used in sound masking systems in hospitals (Ferguson et al., 1997; Mlinek and Pierce, 1997), and studies have demonstrated that it is popular among patients and played a positive role in their well-being, making them feel less tense, more relaxed and safe (Thorgaard et al., 2004). "Day and Night" was favored by most inpatients, with 82% of the patients being very pleased/pleased with the song and 91% of participants defining the sound environment as very pleasant/pleasant (Bitten et al., 2017). Light music without lyrics was selected for this experiment, which avoids experimental deviation caused by the influence of lyrics on patients and their mental health (Baker and Bon, 2008).

The samples used in the experiments were recorded by SQuadriga II with BHS I, and the type of all four sound source samples can be clearly identified. A 5-min sample without dominant sound sources and only ambient noise was recorded as the control group. Five minutes of representative footage from each recording was used as the stimulation material for the experiment, as prolonged use of a VR headset would cause the subjects to become uncomfortable and interfere with the experimental results (Li and Kang, 2019). The 5-min equivalent sound pressure level (SPL) was adjusted to 50dB(A) (Liu and Kang, 2018) for each audio frequency by Audition CS6 to remove differences in volume during the stimulation of the four sounds. To ensure that the participants listened to the four sounds auditory stimuli under similar playback SPL conditions, the LeqA of the audio stimuli had been normalized by an artificial head to 50dB(A) before the experiments to exclude the effect on arousal due to loudness. The background noise was below 45dB(A) during the acoustic stimulation experiment, and nobody spoke in the room. Subjective loudness evaluation were carried out simultaneously, the results show that the loudness level of different groups is significantly different due to the difference of their dominant source frequencies, but the loudness level of different participants in the same group can be ignored.

4.2.2.4 Measurements

By using VR glasses to observe the 3D virtual hospital ward environment created by virtual simulation and headphones to listen to the four types of sound, participants can experience a more realistic hospital environment. The physiological recovery indices include heart rate and skin conductance, which were measured using the

Empatica E4 physiological information monitoring equipment. Information regarding psychological recovery was obtained using a questionnaire. The scale is composed of two parts. The first is psychological feedback, including the anxiety state and perceived environmental restorativeness of the subjects. The anxiety states of participants were measured by using STAI-Y6 with eight questions to indicate their anxiety level (Zijlstra et al., 2017). PRS (the Perceived Restorativeness Scale) was adopted to evaluate the subjects' perceived restorativeness score (Hartig et al., 1997). The second part refers to environmental appraisal. PEQI (the Perceived Environmental Quality Index) was used to describe the participants' perceived environmental quality (Fisher, 1974). The scale consists of a set of bipolar adjectives rated on a seven-point Likert scale, ranging from 1 (extreme negative perception of the environment) to 7 (extreme positive perception of the environment).

4.2.2.5 Procedure

The participants were asked to sit comfortably on a bed. The investigator explained the entire experimental procedure and asked the participants about their physical and mental states. After the subjects understood and agreed to all the terms, the investigator connected the HTC Vive Focus Plus VR device and Empatica E4. The experiment was started after the completion of the connection process and calibration of the physiological signal. The experimental process is shown in Fig. 4.10. First, stress was induced in the subjects using the PASAT (the Packed Audit Serial Addition Task) program. The subjects then received a sound clip (one of the four types of sound sources) and the indoor scene was displayed by the VR equipment. After receiving one experimental condition, the subjects temporarily removed the VR equipment and completed the psychological recovery questionnaire, which took approximately 2min. The subject then rested for 2min, accepted the next sound clip, and followed the process until the four experimental conditions were completed. The sequence of the four experimental conditions in the experiment is followed by a Latin square design (Morsbach et al., 1986). The Empatica E4 equipment continuously recorded the physiological recovery index data of the participants.

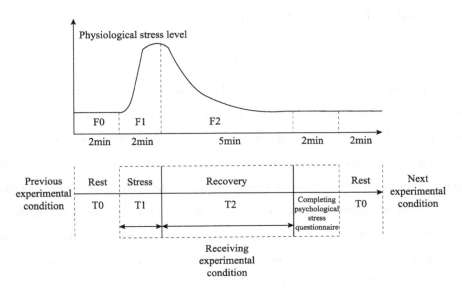

Fig. 4.10 **Experimental process**

4.2.2.6 Data analysis

Regarding physiological data transformation, the heart rate and skin conductance, which are the basic physiological indices of the human body, are easily affected by the physical differences between patients. For example, some individuals may have a relatively high basal heart rate or exhibit more intense physiological responses under the PASAT. Therefore, the physiological stress recovery level of participants cannot be compared by the mean values of physiological indices. In this study, we used the standardized physiological stress recovery rate (R) to estimate the influence of the acoustic environment on the individuals' physiological indices to reduce the potential experimental error caused by physical difference (Payne, 2013; Medvedev et al., 2015; Watts et al., 2016). As shown in formula (1), the R-value can be obtained by dividing the stress recovery level (the difference between the mean values of F1 and F2) by the stress arousal level (the difference between the mean values of F1 and F0). R_{HR} and R_{SCL} represent the stress recovery of the heart rate and skin conductance level, respectively. A higher R-value indicates that, under the experimental conditions, the subjects recover from physiological stress faster.

$$R = \frac{\overline{F_1} - \overline{F_2}}{\overline{F_1} - \overline{F_0}} \qquad (1)$$

For statistical analysis, IBM SPSS 25.0 was used to construct a database containing the final results (Meng et al., 2018; Ba and Kang, 2019). The data were analyzed by the following methods: (1) The differences between the physiological and psychological indicators measured at different time and for different sound source types were determined by repeated analysis of variance measurements, and the level of significance was set at $p < 0.05$. (2) LSD (the least significant difference) post-hoc tests were conducted for pairwise comparisons. The effect sizes (partial η^2) were regarded as minimum, intermediate and high at thresholds of 0.01, 0.06 and 0.14, respectively.

4.2.3 Results

4.2.3.1 Effects of the acoustic environment on physiological stress recovery

The results showed that the patient's mean heart rate recovery rates (R_{HR}) under the ambient sound, mechanical sound, artificial sound and music conditions were 0.66 ($SD = 0.11$), 0.64 ($SD = 0.11$), 0.63 ($SD = 0.12$), and 0.68 ($SD = 0.12$), respectively. As shown in Fig. 4.11, the patients' heart rates tended to recover faster under the music soundscape than the others. However, the repeated measures ANOVA results indicated that the main effect of the soundscape on heart rate recovery was not significant ($F = 1.35$, $p = 0.26$, partial $\eta^2 = 0.04$).

Chapter 4　Sound Environment and Acoustic Perception in Hospitals

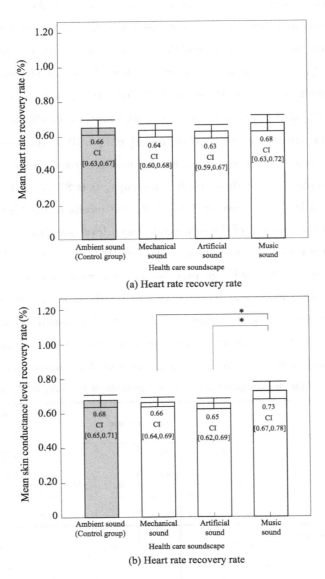

Fig. 4.11　The mean value of patients' physiological stress recovery indicators under exposure to different healthcare soundscape

Note: Error bars represent 95% confidence interval;
* The significant at the 0.05 level.

As assumed, the highest skin conductance level recovery rate (R_{SCL}; $M = 0.73$, $SD = 0.14$) was observed when the patients were exposed to the music soundscape condition. In contrast, the R_{SCL} of patients decreased by 2.5% and 4.6% under the mechanical and artificial sound conditions, respectively, when compared to the control group (ambient sound). The main effect of the soundscape on R_{SCL} was statistically significant ($F = 3.37$, $p = 0.02$), indicating that there was a significant difference in R_{SCL} during exposure to various experimental conditions. Additionally, according to the effect size exhibited by partial η^2 (0.10), the soundscape could exert a substantial impact on R_{SCL}. In the multiple comparisons, a Bonferroni correction is applied. Given there are 6 sub-groups, 0.0083 is used as a corrected significance threshold. As shown in Table 4.8, the results of the LSD post-hoc test confirmed that the recovery of the skin conductance level was faster under the music soundscape than under the mechanical ($MD = 0.07$) and artificial ($MD = 0.08$) sound conditions. Finally, although exposure to mechanical and artificial sound may lead to slower skin conductance level recovery than ambient sound, the difference was not significant. Combining the results of R_{HR} and R_{SCL} partly confirms our first hypothesis; i.e., the healthcare soundscape could impact the physiological stress recovery response of patients, but only in the case of skin conductance level.

Table 4.8 Pairwise comparison of physiological stress recovery levels in patients

Sound type	Heart rate recovery rate				Skin conductance level recovery rate			
	MD	95% CI for Difference		Sig.	MD	95% CI for Difference		Sig.
		Lower Bound	Upper Bound			Lower Bound	Upper Bound	
Ambient sounds-mechanical sounds	0.02	-0.03	0.06	0.48	0.02	-0.02	0.06	0.40
Ambient sounds-artificial sounds	0.03	-0.02	0.07	0.27	0.03	-0.02	0.07	0.24
Ambient sounds-music sound	-0.02	-0.08	0.03	0.42	-0.05	-0.09	0.01	0.07
Mechanical sounds-artificial sounds	0.01	-0.04	0.05	0.70	0.01	-0.03	0.05	0.59

Continued

Sound type	Heart rate recovery rate				Skin conductance level recovery rate			
	MD	95% CI for Difference		Sig.	MD	95% CI for Difference		Sig.
		Lower Bound	Upper Bound			Lower Bound	Upper Bound	
Mechanicalsounds-music sound	−0.04	−0.10	0.02	0.19	−0.07	−0.12	−0.01	0.03 *
Artificial sounds-music sound	−0.05	−0.10	0.01	0.09	−0.08	−0.14	−0.01	0.03 *

Note: * the mean difference is significant at the 0.05 level.

4.2.3.2 Effects of the acoustic environment on psychological stress recovery

The repeated measures ANOVA results suggested that the main effect of healthcare soundscapes on the anxiety state of patients was statistically significant ($F = 10.95$, $p = 0.00$, partial $\eta^2 = 0.26$). Music soundscapes could effectively reduce the patients' anxiety state. After experiencing the music soundscape, the anxiety states of the participants were 16.7%, 14.4% and 24.5% lower than those under the ambient, mechanical and artificial noise conditions, respectively. The LSD post-hoc test indicated that these differences all reached statistical significance ($p < 0.008$). Additionally, artificial noise could cause patients to feel more anxious than the other three soundscapes. As shown in Table 4.9, the LSD post hoc-test revealed that the difference in the anxiety score between mechanical noise and artificial noise was significant ($p = 0.007$), but that between mechanical and control group was not ($p = 0.0027$).

The environmental restorativeness scores given by patients were significantly differed ($F = 9.39$, Sig. $= 0.00$) under the ambient noise, mechanical noise, artificial noise and music soundscape conditions. The effect size (Partial $\eta^2 = 0.23$) suggested the substantial effect of the healthcare acoustic environment on the perceived environmental restorativeness scores. As shown in Fig. 4.12, consistent with the anxiety result, when the music soundscape was broadcast, patients tended to perceive the surrounding environment as "restorative". The percentage of improvement in the restorativeness scores, i, was used to estimate the benefit that certain soundscapes could bring to the participants, which can be calculated as follows: imusic = (PRSmusic-PRSnoise)/PRSnoise. In this formula "imusic" is the percentage of improvement in the restorativeness scores from the music to noise soundscape conditions, "PRSmusic" is the mean perceived restorativeness scores under music soundscape conditions; and "PRSnoise" is the mean perceived restorativeness scores under noise soundscape conditions. The restorativeness scores given to the music condition were 7.9%, 15.0% and 10.5% higher than those given to the ambient, mechanical and artificial noise conditions, respectively. The LSD post-hoc test indicated that the restorativeness differences between music and the other three conditions were all significant ($p < 0.008$). In contrast, participants regarded mechanical noise as the least restorative soundscape ($M = 31.81$, $SD = 4.20$), but the difference between mechanical noise and control group was insignificant.

Table. 4.9 Pairwise comparison of psychological stress recovery levels in patients

Sound type	Heart rate recovery rate				Skin conductance level recovery rate			
	MD	95% CI for difference		Sig.	MD	95% CI for difference		Sig.
		Lower bound	Upper bound			Lower bound	Upper bound	
Ambient sounds -mechanical sounds	0.25	-0.55	1.05	0.530	2.09	0.28	3.91	0.04*
Ambient sounds-artificial sounds	-0.97	-1.82	-0.12	0.027*	0.78	-1.04	2.60	0.35
Ambient sounds -music sound	1.56	0.52	2.61	0.01**	-2.69	-4.51	-0.87	0.00**
Mechanical sounds -artificial sounds	-1.22	-2.08	-0.36	0.01**	-1.32	-3.13	0.51	0.20
Mechanical sounds -music sound	1.31	0.46	2.17	0.00**	-4.78	-6.60	-2.96	0.00**
Artificial sounds -music sound	2.53	1.52	3.54	0.00**	-3.46	-5.29	-1.65	0.00**

Note: * the mean difference is significant at the 0.05 level; ** the mean difference is significant at the 0.01 level; values in bold represent the family-wise error rate corrected significant.

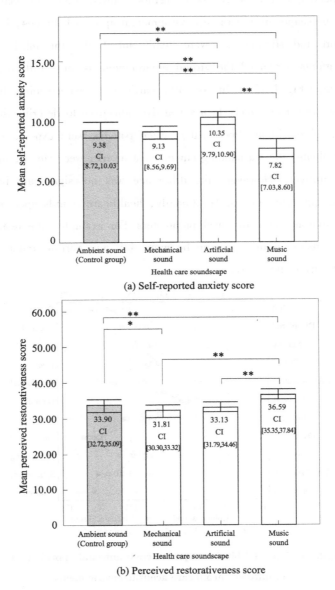

Fig. 4.12 The mean value of patients' psychological stress recovery indicators under exposure to different healthcare soundscape

Note: Error bars depict 95% confidence interval.

* The significant at the 0.05 level. ** The significant at the 0.01 level.

The repeated measures ANOVA results showed that the main effects of healthcare soundscapes on the three environmental appraisal indices, i. e., sense of order, comfort and stimulation, were significant. Under the music experimental condition, patients perceived the virtual environment as more orderly, comfortable and stimulating (Fig. 4.13). In contrast, participants experiencing the mechanical noise condition were more likely to use negative adjectives to describe the sound than under the control conditions. Specifically, when patients were exposed to mechanical noise, they evaluated the surrounding environment as narrower, closed, uncomfortable, artificial and unlively. However, the difference was not significant in any of the environmental appraisal indices. Additionally, healthcare soundscapes may influence some visual environmental appraisal parameters. For example, the results indicated that patients considered the space to be more dull and narrow under artificial and mechanical noise, respectively.

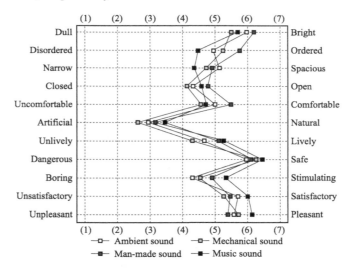

Fig. 4.13 Patients' bipolar adjective environmental appraisal scores in different healthcare acoustic environments

4.2.3.3 Interaction effects of the acoustic environment and demographic factors on stress recovery

As shown in Fig. 4.14, mechanical noise appeared to exert a more negative impact on physiological stress recovery in female patients. Under the mechanical noise condition, the mean R_{HR} and R_{SCL} values of female patients were 1.68% and 3.09% lower than those of male patients. However, the ANOVA results showed that the main effects of gender on R_{HR}, R_{SCL}, anxiety state and perceived restorativeness state were not significant ($p < 0.05$). In addition, there was no significant interaction effect between gender and healthcare soundscape on physiological and psychological stress recovery, indicating that the street recovery outcomes of male and female patients in response to various healthcare soundscapes were similar.

As suggested by Fig. 4.14, patients of different age groups tended to respond to different healthcare soundscapes similarly, although patients aged between 45 and 60 appeared to physiologically recover slightly faster than those in the other age groups. A two-way repeated measures ANOVA was conducted to examine the influence of age on the participants' physiological and psychological stress recovery parameters, and the results indicated that neither the main effect of the patients' age nor the interaction effect between age and soundscape condition was significant ($p < 0.05$), as shown in Table 4.10. The interaction effect between age and soundscape on the perceived restorativeness was almost significant ($p = 0.05$). The senior patients were less sensitive to the three types of healthcare noise, and young people (less than 45 years old) perceived the environment as less restorative under the mechanical noise condition.

The results showed that there was no significant interaction effect between soundscape and demographic characteristics on the participants' restoration; therefore, hypothesis (3) in 4.2.1.2 is rejected. Although we failed to identify any significant interaction effects, certain groups of participants exhibited some particular environmental feedback tendencies. For example, the restorative outcome of elderly people appeared to be less sensitive to acoustical conditions. Participants aged between 45 and 60 years tended to withstand negative sounds better than those in the other age groups. In future studies, we could consider the patients; socioeconomic characteristics in analysis to explore the potential interaction effects.

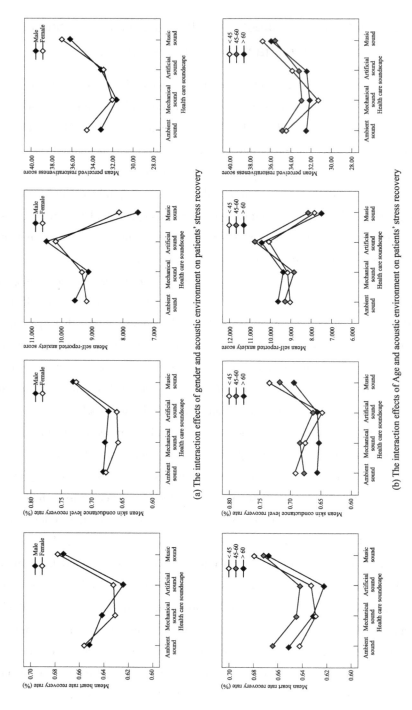

Fig. 4.14 The interaction effects and acoustic environment on patients' stress recovery

Table. 4.10 ANOVA results for main and interaction effects on physiological and psychological recovery

Influence factors	Physiological stress recovery				Psychological stress recovery			
	R_{HR}		R_{SCL}		STAI		PRS	
	F	Sig.	F	Sig.	F	Sig.	F	Sig.
Gender	0.01	0.94	0.30	0.59	0.01	0.91	0.91	0.34
Gender with sound type	0.04	0.99	0.05	0.99	0.57	0.64	0.26	0.86
Age	0.23	0.79	2.00	0.19	0.28	0.73	0.82	0.45
Age with sound type	0.36	0.91	1.07	0.36	1.78	0.10	2.14	0.05

4.2.4 Discussion

The results obtained from the skin conductance level partly support the theory that the healthcare soundscape could affect physiological stress recovery. The music soundscape plays a restorative effect on healthcare. The result showned the recovery rate of participants' SCL under the music condition was faster than that under the ambient, mechanic and artificial sound conditions. However, none of these differences reached significant, indicating that the restorative effects of music are limited in the aspect of physiological stress.

Some previous studies observed a stronger restorative effect of music than this study (Medvedev et al., 2015). The experimental setting we adopted may cause different results, as VR conditions could draw the patients' attention to visual stimuli and weaken the restorative effect of soundscapes. Another possible explanation may be that the music used in this study was not self-selected. Researchers have confirmed that a participant's sense of control could improve restorative effects (Heitz et al., 1992). Additionally, the different musical tastes of patients could affect the restorative impact (Watts et al., 2016).

No significant impact of soundscape on the patients' R_{HR} was observed here, which is consistent with the results of some previous studies (Aletta et al., 2018a; Yu et al., 2018). This may be attributed to the special characteristics of heart rate, which is highly sensitive to the mode of information processing (Ulrich et al., 1991;

Meehan et al. , 2005). A person's heart rate accelerates dramatically if experimental conditions involve information processing, such as mental counting (Lacey and Lacey, 1974). This study involved no information storage, retrieval or manipulation under any of the four soundscapes, which may have resulted in an insignificant effect on heart rate. Additionally, although the skin conductance level and heart rate are both indicators of sympathetic nervous system activity, the sensitivities when testing various built environmental stimuli may differ (Cacioppo et al. , 2007). For example, studies have found that skin conductance is sensitive to changes in the luminous environment (Izso et al. , 2009), and heart rate is more responsive to the facade pattern (Chamilothori et al. , 2019).

The second hypothesis was confirmed by the significant effects of soundscape on the participants' anxiety level and perceived restorativeness state. Artificial induced higher anxiety than the mechanical experimental conditions in the study. A potential explanation for this result is that artificial sound contains more transient noise (Allaouchiche et al. , 2002), which may negatively impact psychological recovery. However, the anxiety level of the participants experiencing the mechanical sound was not significantly higher than that of the participants in the control group. Although this is the first study in this area, the influence of the participants' acoustic expectations may explain this result. The potential function of space (such as socializing and working) could influence a user's environmental expectations and evaluations (Chamilothori et al. , 2019). In this study, healthcare was chosen as the research context, and participants may anticipate certain kinds of mechanical noise before being subjected to the experimental conditions. Thus, its negative impact on recovery could be relieved.

The perceived restorativeness state was also significantly influenced by healthcare soundscapes. Mechanical noise was perceived as the least restorative condition in the study, which was inconsistent with the anxiety state data. This might be due to the different assessment weights between the two psychological recovery indicators. The anxiety state assesses the participants' mental state. For example, the scale we used included items such as "I feel upset" (Marteau and Bekker, 1992). However, the perceived restorativeness state reflects the external environment appraisal. In this study, both parameters may partly reflect the impact of the soundscape on the patients' psychological states, but more work should be conducted to determine the mechanisms and pathways of the effects of healthcare soundscapes on psychological stress recovery.

This research assessed the effects of the acoustic environment on patients' environmental appraisal in a healthcare facility, and the data show that healthcare soundscapes may have affected the participants' environmental appraisal. The music condition was perceived as being more positive than the three other experimental conditions for nine of the eleven evaluation dimensions, and could significantly improve the patients' environmental perception in terms of the order, comfort and stimulation. However, no significant difference was observed between the evaluation of mechanical, artificial and ambient noise. Overall, the soundscape has less of an impact on patients' environmental appraisal than visual information, such as color, lighting and spatial layout (Leather et al., 2003). The smaller effect may indicate that visual stimulation is a dominating factor affecting environmental appraisal in healthcare settings.

This study found that the soundscape could alter the patients' visual impression of the environment, such as their sense of light, order and scale. This could be because people holistically perceive the environment, and audio and visual stimuli could drive multisensory environmental perception (Viollon et al., 2002; Pheasant et al., 2010). Attractive or meaningful visual contexts tend to increase people's tolerance to noise, causing less irritation in similar noisy acoustic environments (Bangjun et al., 2003; Iachini et al., 2012). However, studies have also observed a high correlation between audio information and an individual's visual experience and preference. In this study, the participants tended to regard the surrounding environment as orderly, comfortable and stimulating under the music condition, which may be because the sound stimuli altered the participants' visual cognitive processing (Ren and Kang, 2015).

Generally, the objective data were relatively consistent with the subjective ratings (anxiety, perceived restorativeness state and environmental appraisal), which could verify the validity of the method used in the study. However, when faced with audio stimuli, the psychological stress recovery indicator tended to be more sensitive than physiological parameters. The effect size also indicates that the soundscape could exert a greater effect on the outcome of psychological factors, supporting the results of previous studies. This may be because physiological stress recovery parameters, such as heart rate, skin conduction levels or blood pressure, are indices of sympathetic arousal and cannot reflect the valence of emotion (Ward and Marsden, 2003). Therefore, physiological data cannot detect mild arousal responses coupled with positive emotional reactions. Therefore, interpreting a person's stress level using physiological data alone is insufficient, especially considering the stress reaction, and recovery is a complex process involving cognition and reflection (Bartlett, 1998).

4.2.5 Conclusion

Inpatients are more prone to anxiety and stress than healthy individuals; therefore, hospital wards must provide suitable acoustic environments to help them to relax and recover. The restorative effects of soundscapes have been investigated, but few studies have been conducted on patients and hospital environments. This study mainly explores and analyses the influence of the acoustic environment on physiological/psychological stress indicators of patients in hospital wards.

The impact of soundscape on patients' physiological stress parameters was relatively modest. In this study, sounds did not significantly impact the patients' heart rate recovery rates (R_{HR}). However, the results demonstrate that the soundscape could significantly influence the patients' skin conductance level recovery rate (R_{SCL}). The recovery rate was faster under music than the mechanical or artificial noise conditions, though the difference fails to reach significantly.

The acoustic environment could exert profound effects on the patients' psychological stress indicators, with both the patients' self-reported anxiety state and perceived restorativeness score significantly affected by the healthcare soundscapes. Patients continuously reported less anxiety and higher perceived restorativeness for the music soundscape than the ambient, mechanical and artificial noise soundscapes. The reported anxiety levels were highest under the artificial sounds and mechanical sounds were regarded as the least restorative. For the environmental appraisal of psychological parameters, the music condition was described as significantly more ordered, comfortable and stimulating than the three other experimental conditions. There was no statistically significant difference between the environmental appraisal of mechanical, artificial and ambient sounds. However, it was found that the acoustic environment could alter the patients' visual impression of the environment.

The interaction effects of gender, age and acoustic environment were not significant. However, there were some environmental feedback tendencies for certain groups of participants, and future studies may consider the patients' social-economic characteristics. Hospital spaces are rather diverse; thus, it would be interesting to consider other spaces, such as outpatient halls, waiting rooms, double beds and dormitory bed wards, in future studies. While this study indicated that the acoustic environment of hospital wards influences the physiological and psychological indices of patients, and also demonstrated that virtual reality is an effective method of analyzing the relative influences of different dominant sound sources; in future work, the absolute influence of acoustic environment on the psychological and physiological indicators of patients could be examined in realistic environments.

References

[1] ALETTA F, AXELSSON, KANG J. Dimensions underlying the perceived similarity of acoustic environments [J]. Frontiers in psychology, 2017, 8: 1162.

[2] ALETTA F, KANG J, AXELSSON Ö. Soundscape descriptors and a conceptual framework for developing predictive soundscape models [J]. Landscape and Urban Planning, 2016, 149: 65-74.

[3] ALETTA F, MASULLO M, MAFFEI L, et al. The effect of vision on the perception of the noise produced by a chiller in a common living environment [J]. Noise Control Engineering Journal, 2016, 64 (3): 363-378.

[4] ALETTA F, OBERMAN T, KANG J. Associations between positive health-related effects and soundscapes perceptual constructs: A systematic review [J]. International journal of environmental research and public health, 2018, 15 (11): 2392.

[5] ALETTA F, OBERMAN T, KANG J. Positive health-related effects of perceiving urban soundscapes: a systematic review. The Lancet, 2018, 392 (2): S3.

[6] ALETTA F, VAN RENTERGHEM T, BOTTELDOOREN D. Influence of personal factors on sound perception and overall experience in urban green areas. A case study of a cycling path highly exposed to road traffic noise [J]. International journal of environmental research and public health, 2018, 15 (6): 1118.

[7] ALLAOUCHICHE B, DUFLO F, DEBON R, et al. Noise in the postanaesthesia care unit [J]. British Journal of Anaesthesia, 2002, 88 (3): 369-373.

[8] ALVARSSON J J, WIENS S, NILSSON M E. Stress recovery during exposure to nature sound and environmental noise [J]. International journal of environmental research and public health, 2010, 7 (3): 1036-1046.

[9] ANNERSTEDT M, JÖNSSON P, WALLERGÅRD M, et al. Inducing physiological stress recovery with sounds of nature in a virtual reality forest—Results from a pilot study [J]. Physiology & Behavior, 2013, 118 (13): 240-250.

[10] BA M, KANG J. Effect of a fragrant tree on the perception of traffic noise [J]. Building and Environment, 2019, 156: 147-155.

[11] BAKER C F. Discomfort to environmental noise: Heart rate responses of SICU patients [J]. Critical Care Nursing Quarterly, 1992, 15 (2): 75.

[12] BAKER F, BOR W. Can music preference indicate mental health status in young people? [J]. Australasian Psychiatry, 2008, 16 (4): 284-288.

[13] BANGJUN Z, LILI S, GUOQING D. The influence of the visibility of the source on the subjective annoyance due to its noise [J]. Applied Acoustics, 2003, 64 (12): 1205-1215.

[14] BARRY M, FRIEDLI L. The influence of social, demographic and physical factors on positive mental health in children, adults and older people [M]. In: Mental Capital and Wellbeing. Wiley-Blackwell, 2010: 475-484.

[15] BARTLETT D. Stress: Perspectives and processes [M]. Buckingham, UK: McGraw-Hill Education, 1998.

[16] BELOJEVIC G, PAUNOVIC K, JAKOVLJEVIC B, et al. Cardiovascular effects of environmental noise: research in Serbia [J]. Noise and Health, 2011, 13 (52): 217.

[17] BERGLUND B, LINDVALL T, SCHWELA D H, et al. Guidelines for community noise. World Health Organization: Protection of the Human Environment, 1999.

[18] BEUTEL M E, JÜNGER C, KLEIN E M, et al. Noise annoyance is associated with depression and anxiety in the general population-the contribution of aircraft noise [J]. Plos one, 2016, 11 (5): e0155357.

[19] BITTEN T, BIRGITTE B H, GUNHILD P, et al. Specially selected music in the cardiac laboratory—an important tool for improvement of the wellbeing of patients [J]. European

Journal of Cardiovascular Nursing, 2017, 3 (1): 21-26.

[20] BUCKLES E. Evaluation of patient satisfaction in A&E [J]. Nursing standard (Royal College of Nursing (Great Britain), 1990, 4 (19): 33-35.

[21] CACIOPPO J T, TASSINARY L G, BERNTSON G. Handbook of psychophysiology [M]. Cambridge, UK: Cambridge university press, 2007.

[22] CHAMILOTHORI K, CHINAZZO G, RODRIGUES J, et al. Subjective and physiological responses to Subjective and physiological responses to façade and sunlight pattern geometry in virtual reality and sunlight pattern geometry in virtual reality [J]. Building and Environment, 2019, 150: 144-155.

[23] CHAMILOTHORI K, WIENOLD J, ANDERSEN M. Adequacy of immersive virtual reality for the perception of daylit spaces: comparison of real and virtual environments [J]. Leukos, 2019, 15 (1-4), 203-226.

[24] CollignonO, Girard S, Gosselin F, et al. Audio-visual integration of emotion expression. Brain Research, 2008, 1242: 126-135.

[25] DEROY O, SPENCE C. Crossmodal correspondences: four challenges [J]. Multisensory Research, 2016, 29 (1-3): 29-48.

[26] EISENMANN A. Verkehrslärm, lärmbezogene Gestörtheit, kognitive Leistung, Schlafqualität und Wohlbefinden bei Volksschulkindern im alpinen Raum. na. , 2006.

[27] EVANS G W. The built environment and mental health. Journal of urban health, 2003, 80 (4): 536-555.

[28] FARREHI P M, NALLAMOTHU B K, NAVVAB M. Reducing hospital noise with sound acoustic panels and diffusion: a controlled study [J]. Bmj Quality & Safety, 2016, 25 (8): 644-646.

[29] FERGUSON E, SINGH A P, CUNNINGHAM-SNELL N. Stress and blood donation: effects of music and previous donation experience. British Journal of Psychology, 1997, 88 (Pt 2): 277-294.

[30] FISHER J D. Situation specific variables as determinants of perceived environmental aestheticquality and perceived crowdedness [J]. Journal of Research in Personality, 1974, 8 (2): 177-188.

[31] HAMMERSEN F, NIEMANN H, HOEBEL J. Environmental noise annoyance and mental health in adults: findings from the cross-sectional German Health Update (GEDA) Study 2012 [J]. International journal of environmental research and public health, 2016, 13 (10): 954.

[32] HAMILTON-FLETCHER G, WARD J, WRIGHT T D. Cross-modal correspondences

enhance performance on a colour-to-sound sensory substitution device [J]. Multisensory Research, 2016, 29 (4-5): 337-363.

[33] HARTIG T, KORPELA K, EVANS G W, et al. A measure of restorative quality in environments [J]. Scandinavian housing and planning research, 1997, 14 (4): 175-194.

[34] HARTIG T, STAATS H. Guest Editors' introduction: Restorative environments [J]. Journal of Environmental Psychology, 2003, 23 (2): 103-107.

[35] HEITZ L, SYMRENG T, SCAMMAN F. Effect of music therapy in the postanesthesia care unit: a nursing intervention [J]. Journal of Post Anesthesia Nursing, 1992, 7 (1): 22-31.

[36] HEYDARIAN A, CARNEIRO J P, GERBER D, et al. Immersive virtual environments versus physical built environments: a benchmarking study for building design and user-built environment explorations [J]. Automation in construction, 2015, 54: 116-126.

[37] HUME K, AHTAMAD M. Physiological responses to and subjective estimates of soundscape elements [J]. Applied Acoustics, 2013, 74 (2): 275-281.

[38] STAMPSIII E. Evaluating spaciousness in static and dynamic media [J]. Design Studies, 2007, 28 (5): 535-557.

[39] IZSO L, LÁNG E, LAUFER L, et al. Psychophysiological, performance and subjective correlates of different lighting conditions [J]. Lighting Research & Technology, 2009, 41 (4): 349-360.

[40] JENSEN H A, RASMUSSEN B, EKHOLM O. Neighbour and traffic noise annoyance: a nationwide study of associated mental health and perceived stress [J]. European journal of public health, 2018, 28 (6): 1050-1055.

[41] JOYNT J L, KANG, J. The influence of preconceptions on perceived sound reduction by environmental noise barriers [J]. Science of the total environment, 2010, 408 (20): 4368-4375.

[42] KANG J, ALETTA F, GJESTLAND T T, et al. Ten questions on the soundscapes of the built environment [J]. Building and environment, 2016, 108: 284-294.

[43] KAPLAN R, KAPLAN S. The experience of nature: A psychological perspective [M]. New York, NY, US: Cambridge University Press, 1989.

[44] KNFERLE K, SPENCE C. Crossmodal correspondences between sounds and tastes. Psychonomic Bulletin & Review, 2012, 19 (6): 1-15.

[45] LACEY B C, LACEY J I. Studies of heart rate and other bodily processes in sensorimotor behavior [M]. In: Cardiovascular psychophysiology: Current issues in response mechanisms,

biofeedback and methodology. New Brunswick, NJ, US: AldineTransaction, 1974: 538-564.

[46] IACHINI T, MAFFEI L, RUOTOLO F, et al. Multisensory assessment of acoustic comfort aboard metros: a virtual reality study [J]. Applied Cognitive Psychology, 2012, 26 (5): 757-767.

[47] LEATHER P, BEALE D, SANTOS A, et al. Outcomes of environmental appraisal of different hospital waiting areas [J]. Environment and Behavior, 2003, 35 (6), 842-869.

[48] LI Z, KANG J. Sensitivity analysis of changes in human physiological indicators observed in soundscapes [J]. Landscape and Urban Planning, 2019, 190: 103593.

[49] LIU F, KANG J. Relationship between street scale and subjective assessment of audio-visual environment comfort based on 3D virtual reality and dual-channel acoustic tests [J]. Building and Environment, 2018, 129: 35-45.

[50] MACKENZIE D, GALBRUN L. Noise levels and noise sources in acute care hospital wards [J]. Building Services Engineering Research and Technology, 2007, 28 (2): 117-131.

[51] MARTEAU T M, BEKKER H. The development of a six - item short - form of the state scale of the Spielberger State—Trait Anxiety Inventory (STAI). British journal of clinical Psychology, 1992, 31 (3): 301-306.

[52] MEDVEDEV O, SHEPHERD D, HAUTUS M J. The restorative potential of soundscapes: A physiological investigation [J]. Applied Acoustics, 2015, 96: 20-26.

[53] MEEHAN M, RAZZAQUE S, INSKO B, et al. Review of four studies on the use of physiological reaction as a measure of presence in stressful virtual environments [J]. Applied Psychophysiology and Biofeedback, 2005, 30 (3): 239-258.

[54] MENG Q, ZHAO T, KANG J. Influence of music on the behaviors of crowd in urban open public spaces [J]. Frontiers in psychology, 2018, 9: 596.

[55] MEYER K, KAPLAN J T, ESSEX R, et al. Predicting visual stimuli on the basis of activity in auditory cortices [J]. Nature neuroscience, 2010, 13 (6): 667.

[56] MLINEK E J, PIERCE J. Confidentiality and privacy breaches in a university hospital emergency department [J]. Academic Emergency Medicine, 1997, 4 (12): 1142-1146.

[57] MORSBACH G, MCCULLOCH M, CLARK A. Infant crying as a potential stressor concerning mother' concentration ability [J]. Psychologia: An International Journal of Psychology in the Orient, 1986, 29 (1): 18-20.

[58] OKOKON E O, TURUNEN A W, UNG-LANKI S, et al. Road-traffic noise: annoyance, risk perception, and noise sensitivity in the Finnish adult population [J]. International journal of environmental research and public health, 2015, 12 (6): 5712-5734.

[59] OZTURK O, KREHM M, VOULOUMANOS A. Sound symbolism in infancy: evidence for sound-shape cross-modal correspondences in 4-month-olds [J]. Journal of Experimental Child Psychology, 2013, 114 (2): 173-186.

[60] PAUNOVIĆ K, STOJANOV V, JAKOVLJEVÍ B, et al. Thoracic bioelectrical impedance assessment of the hemodynamic reactions to recorded road-traffic noise in young adults [J]. Environmental research, 2014, 129: 52-58.

[61] PAYNE S R. The production of a Perceived Restorativeness Soundscape Scale [J]. Applied Acoustics, 2013, 74 (2): 255-263.

[62] PHEASANT R J, FISHER M N, WATTS G R, et al. The importance of auditory-visual interaction in the construction of 'tranquil space' [J]. Journal of environmental psychology, 2010, 30 (4): 501-509.

[63] QIN X, KANG J, JIN H. Subjective evaluation of acoustic environment of waiting areas in general hospitals [J]. Building Science, 2011, 12: 53-60.

[64] RASHID M, ZIMRING C. A review of the empirical literature on the relationships between indoor environment and stress in health care and office settings: Problems and prospects of sharing evidence [J]. Environment and Behavior, 2008, 40 (2): 151-190.

[65] RECIO A, LINARES C, BANEGAS J R, et al. Road traffic noise effects on cardiovascular, respiratory, and metabolic health: An integrative model of biological mechanisms [J]. Environmental research, 2016, 146: 359-370.

[66] REN X, KANG J. Interactions between landscape elements and tranquility evaluation based on eye tracking experiments [J]. The Journal of the Acoustical Society of America, 2015, 138 (5): 3019-3022.

[67] SHEPLEY M M, BAUM M, GINSBERG R, et al. Eco-effective design and evidence-based design: Perceived synergy and conflict [J]. HERD (Health Environments Research & Design Journal), 2009, 2 (3): 56-70.

[68] SPENCE C. Crossmodal correspondences: a tutorial review [J]. Attention, perception & psychophysics, 2011, 73 (4), 971-995.

[69] TAJADURA-JIMÉNEZ A, LARSSON P, VÄLJAMÄE A, et al. When room size matters: acoustic influences on emotional responses to sounds [J]. Emotion, 2010, 10 (3): 416.

[70] THORGAARD B, HENRIKSEN B B, PEDERSBAEK G, et al. Specially selected music in the cardiac laboratory—an important tool for improvement of the wellbeing of patients [J]. European Journal of Cardiovascular Nursing, 2004, 3 (1): 21-26.

[71] TOBÍAS A, RECIO A, DÍAZ J, et al. Noise levels and cardiovascular mortality: a case-crossover analysis [J]. European journal of preventive cardiology, 2015, 22 (4):

496-502.

[72] ULRICH R S, SIMONS R F, LOSITO B D, et al. Stress recovery during exposure to natural and urban environments [J]. Journal of environmental psychology, 1991, 11 (3): 201-230.

[73] ULRICH R S, ZIMRING C, ZHU X, et al. A review of the research literature on evidence-based healthcare design [J]. HERD (Health Environments Research & Design Journal), 2008, 1 (3): 61-125.

[74] VIOLLON, S, LAVANDIER, C, AND DRAKE, C (2002). Influence of visual setting on sound ratings in an urban environment. Applied acoustics 63 (5), 493-511.

[75] VON LINDERN E, HARTIG T, LERCHER P. Traffic-related exposures, constrained restoration, and health in the residential context [J]. Health & place, 2016, 39: 92-100.

[76] WARD R D, MARSDEN P H. Physiological responses to different WEB page designs [J]. International Journal of Human-Computer Studies, 2003, 59 (1-2): 199-212.

[77] WATTS G, KHAN A, PHEASANT R. Influence of soundscape and interior design on anxiety and perceived tranquillity of patients in a healthcare setting [J]. Applied Acoustics, 2016, 104: 135-141.

[78] WU Y, MENG Q, LI L, et al. Interaction between sound and thermal influences on patient comfort in the hospitals of China's northern heating region [J]. Applied Sciences, 2019, 9 (24): 5551.

[79] XIE H, KANG J. The acoustic environment of intensive care wards based on long period nocturnal measurements [J]. Noise and Health, 2012, 14 (60): 230-236.

[80] XIE H, KANG J. Sound field of typical single-bed hospital wards [J]. Applied Acoustics, 2012, 73 (9): 884-892.

[81] XIE H, KANG J, MILLS G H. Clinical review: The impact of noise on patient' sleep and the effectiveness of noise reduction strategies in intensive care units [J]. Critical Care, 2009, 13 (2): 208.

[82] XIE H, KANG J, MILLS G H. Behavior observation of major noise sources in critical care wards [J]. Journal of Critical Care, 2013, 28 (6): 1109. e5-1109. e18.

[83] YIN J, ZHU S, MACNAUGHTON P, et al. Physiological and cognitive performance of exposure to biophilic indoor environment [J]. Building and Environment, 2018, 132: 255-262.

[84] YOST W, ZHONG X. Localizing sound sources when the listener moves: Vision required [J]. The Journal of the Acoustical Society of America, 2015, 137 (4), 2373-2373.

[85] YU C-P, LEE H-Y, LUO X-Y. The effect of virtual reality forest and urban environments on physiological and psychological responses [J]. Urban Forestry & Urban Greening, 2018, 35: 106-114.

[86] ZIJLSTRA E, HAGEDOORN M, KRIJNEN W P, et al. Motion nature projection reduces patient's psycho-physiological anxiety during ct imaging [J]. Journal of Environmental Psychology, 2017, 53: 168-176.

[87] ZHANG X, LIAN Z, DING Q. Investigation variance in human psychological responses to wooden indoor environments [J]. Building and Environment, 2016, 109: 58-67.

[88] ZWICKER E, FASTL H. Psychoacoustics: Facts and models [J]. Physics Today, 1999, 54 (6): 64-65.

Chapter 5

Sound Environment and Acoustic Perception in Living Spaces

5.1 Effects of sound environment on the sleep of college students

5.1.1 Introduction

A report from China's Ministry of Education noted that about 31.43 million students live in residence halls, accounting for 82% college students in China, and 80% of dorms are four-person or even six-person bedrooms (Ministry of Education of the People's Republic of China, 2019). Sleep plays an essential role in college students' daily life (Gawlik et al., 2019; George, 2007; Gregory et al., 2004; Orzel-Gryglewska, 2010; Pilcher and Huffcutt, 1996; Zaharna and Guilleminault, 2010). Such a complex and unique environment inevitably brings about an impact on sleep for college students. First, sleep is the only other biological necessity of our bodies in addition to air, water and food (Gregory et al., 2004). Sleep loss has been implicated in a variety of adverse health outcomes (Zaharna and Guilleminault, 2010), including cardiovascular abnormalities (Gawlik et al., 2019), immunological problems (Orzeł-Gryglewska, 2010), psychological health concerns (Pilcher and Huffcutt, 1996), and neuro-behavioural impairment that can lead to accidents (George, 2007). Furthermore, college years are when students gain critical knowledge, skills, human capital and credentials to become successfully employed and contribute to society after graduation (Astin et al., 2019). Previous studies have confirmed that sleep-deprived students tend to learn material less efficiently, which could lead to a lower grade point average (GPA; Gilbert and Weaver, 2010). When sleep deprivation occurs during college years, it presents an obstacle to maximising individuals' success during this critical time.

There have been studies that have focused on sleep among college students (Hershner and Chervin, 2014; Lund et al., 2010; Van Dongen et al., 2003). However, some studies have reported high rates of insufficient sleep among college students (Gilbert and Weaver, 2010; Lund et al., 2010). Therefore, a better understanding of the causes of sleep deficits in the college student population is needed. Sleep deprivation occurs for multiple reasons, some of which are physiological and others behavioural. Many students have inadequate sleep hygiene that, in conjunction with their delayed circadian rhythm, contributes to sleep deprivation (Hershner and Chervin, 2014). The mechanisms of some influential factors have been studied. In terms of substance use, alcohol use has been shown to lead to shortened sleep latency and the promotion of fragmented sleep in the latter half of the night (Amlander and Fuller, 2005). Caffeine, taken at night, has been found to increase sleep latency, reduce sleepiness, and improve the ability to sustain wakefulness (Walsh et al., 1990). Further, stimulants have been demonstrated to increase sleep latency, suppress rapid eye movement sleep (REM sleep), and worsen sleep quality (Amlander and Fuller, 2005; Clegg-Kraynok et al., 2011). Frequent cell phone use at bedtime has been associated with difficulties falling asleep, repeated awakenings, or waking up too early (Thomée et al., 2007). Additionally, researches have indicated that noise is a significant sleep disturbance in college residential sleep environments (Lund et al., 2010; Sexton-Radek and Hartley, 2013). Such studies on the factors influencing sleep have rarely considered how nocturnal environmental noise might affect sleep among college students living in residence halls. However, the sound environments of school gymnasia (Maffei et al., 2009), offices (Kang et al., 2017) and classrooms (Ricciardi and Buratti, 2018) have been studied; therefore, more in-depth research on factors in residence halls that affect college students' sleep is needed.

Sleep has been shown to be disturbed by sound environment (Basner et al., 2010, 2011), and when this disturbance is severe and frequent, it can lead to significant sleep fragmentation and sleep deprivation, which is detrimental to physical and mental health (Basner et al., 2010). Nocturnal environmental noise is considered a major cause of exogenous sleep disturbance, after somatic problems and day tensions (Basner et al., 2011). Most previous studies focused on a certain kind of noise or a specific group of people and have shown that residents exposed to nocturnal environmental noise such as aircraft (Kwak et al., 2016), road (Pirrera et al., 2014), train (Smith et al., 2016), wind turbine noises (Bakker et al., 2012) or the combination of noises (Lee et al., 2010) at night exceeding a certain sound level can result in sleep disturbances. The effects of noise on sleep among different populations, including infants (Strauch et al., 1993), children (Weyde et al., 2017), adults (Evandt et al., 2017), and other special populations, such as hospitalised patients (Park et al., 2014) and some categories of workers (Azadboni et al., 2018) have been studied.

However, relatively less research has examined how nocturnal environmental noise affects college students' sleep in communal living environments that are primarily occupied by four to six persons. When sharing a small room, occupants are exposed to noises from various activities. Research that explores the effects of sound environment on sleep among college students in China may help identify ways to improve their sleep quality. Therefore, using a typical university residence hall in Harbin city, China as a case site, and using subjective and objective measures, the present study examined four research questions. First, we addressed the question regarding sound environment being one of the main factors contributing to disturbed sleep in college students, and if it was, what is the proportion and extent of the students' sleep disturbance? Second, we asked, how does the number of occupants affect the nocturnal sound environment of residence halls? Third, we explored how sound levels affect the sleep quality of college students. Fourth, we examined how sound sources affect the sleep quality of college students.

5.1.2 Materials and methods

5.1.2.1 Case site

Most Chinese university dorms are multi-storey buildings with five to seven floors. Generally, they are planned on a campus, away from outdoor noise resulting from road traffic and industry, and separate units are created for men and women. In China, about 65% of college dormitories are four-person rooms; 16% are six-person rooms; and the remaining 19% are single rooms, double rooms and eight-person rooms. The living area of these units is commonly between 16 and 18 square metres. In addition, undergraduate and graduate students are the primary groups in Chinese colleges, accounting for 98.8% of the nation's college students (Ministry of Education of the People's Republic of China, 2019).

Field surveys are essential to evaluate the practical environmental settings for sleep (Stansfeld et al., 2000). Since noise is a common environmental hazard, it is a simpler approach to explore the effects of noise in everyday life rather than in laboratory experiments. Thus, field studies are more accurate in reflecting participants' actual sleep patterns (Öhrström and Skanberg, 2004). Study locations were drawn from areas at Harbin Institute of Technology in north-eastern China. It has about 46,000 students recruited from all over the country, which make the research results representative of the situation of college students nationwide (Harbin Institute of Technology, 2019). A typical residence hall was chosen for this field survey considering that the focus of the study was the impact of different buildings and their surroundings. The dormitory accommodates 1,262 students, including both undergraduate and graduate students, with approximately equal numbers of men and women. Only four-person and six-person bedrooms are present with an area of 17.5 square metres, and the number of occupants varies from one to six. The building is seven floors high and adjacent to three streets, with one side facing the courtyard. As shown in Fig. 5.1, the other sides face Lianfa Street, Haicheng Street and Gongsi Street, all pedestrian streets with a small amount of traffic. Therefore, there are various typical sound sources inside the dormitory building, including snoring, conversation, whispering, footsteps and keyboard/mouse click with a little outdoor noise.

Chapter 5 Sound Environment and Acoustic Perception in Living Spaces

Fig. 5. 1 **Plan of survey site**

Male students were living on the first to the third floors, and female students were living on the fifth to seventh floors. The fourth floor is divided into two separate parts for women and men. As shown in Fig. 5. 2, bedrooms of equal size are on both sides of the corridor. The bedroom is 5 metres long and 3. 5 metres wide, and there are two dwelling modes. Students of different grades are distributed on every floor. The undergraduate students primarily reside in six-person bedrooms, which includes upper and lower bunks as shown in Fig. 5. 3 (a), while the rooms for graduate students are primarily four-person bedrooms, and loft bed with a desk is the main mode as shown in Fig. 5. 3 (b). Sometimes the number of people living in the bedroom was fewer than the standard number, due to arrangement by the university or personal reasons. Therefore, the number of occupants in the rooms varied from one to six.

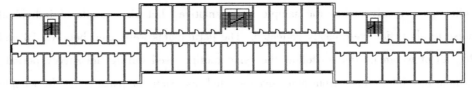

Fig. 5. 2 **Floor plan of the residence hall**

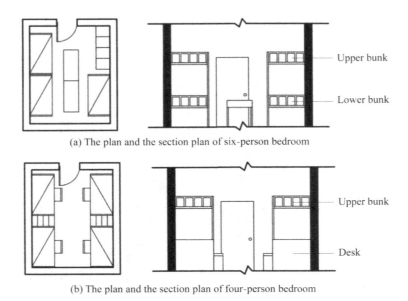

(a) The plan and the section plan of six-person bedroom

(b) The plan and the section plan of four-person bedroom

Fig. 5.3 The plan and the section plan of bedroom

5.1.2.2 Participants

Prior studies have indicated that 80-100 participants are enough for a questionnaire study on indoor noise and sleep (Ising et al., 2002; Lercher et al., 2010; Thomas et al., 2012). Therefore, considering the variety of gender and grade of students, the number of occupants, as well as different floors, 105 students in the residence hall were chosen for the present study with stratified sampling. They were informed of the purpose, procedures and requirements of the study through an e-mail and asked to provide informed consent by e-mail reply. Of them, 93 were willing to participate in the survey, including three participants who did not live in the residence hall during the survey. 90 valid responses were received. Such sample collection not only makes the sample representative, but also details of individual noise levels inside were easily collected. The participants were aged between 18 and 26, all of whom needed to follow the same routine used during the survey period, without any interference from other events, such as exams or parties. Of the sample 50% ($n=45$) were men, and 50% ($n=45$) were women. Of these, 15.6% were freshmen ($n = 14$), 15.6% were sophomores ($n = 14$), 17.8% were juniors ($n = 16$), 17.8% were seniors ($n = 16$), 17.8% were first-year graduates ($n = 16$), and 15.6% were second-year graduates ($n = 14$). The mean age of the sample was 21 years ($SD = 1.9$). This study was reviewed and approved by the university's institutional review board. Written informed consent was obtained from all participants. As incentives for participation, participants received either a monetary reward or various learning materials reward.

5.1.2.3 Subjective measurements

Participants were required to complete a questionnaire once to collect long-term evaluations that included the Pittsburgh Sleep Quality Index (PSQI), other personal characteristics, residence information, and subjective assessment of the nocturnal sound environment.

1. Demographic survey

Participants completed a brief demographic survey. This survey included questions about age, gender, grade level, height, weight and the number of occupants per room. The position of the bed, including upper bunk beside the window, upper bunk beside the door, lower bunk beside the window, and lower bunk beside the door, were also recorded.

2. Self-reported sleep quality

PSQI is a measure of the subjective experience of sleep that is a well-established psychometric assessment (Carpenter and Andrykowski, 1998) and is one of the most common subjective methodologies used in sleep research. It can provide an assessment of sleep quality over the previous 30 days. PSQI differentiates between poor-quality and good-quality sleepers by measuring seven areas: subjective sleep quality, sleep latency, sleep duration, habitual sleep efficiency, sleep disturbances, use of sleep medication, and daytime dysfunction over the past month (Buysse et al., 1989). The components of PSQI are scored on a scale from 0 (better) to 3 (worse); therefore, the global PSQI is a score ranging from 0-21. A global score (>5) indicates a poor-quality sleep, whereas a score of 5 is indicative of a good-quality sleep (Spira et al., 2012). For this study, the global PSQI scores were split into three categories: optimal (≤ 5), borderline (6-7), and poor (≥ 8) sleep quality for the purpose of achieving relatively even group sizes (Lund et al., 2010). The internal consistency of PSQI in this study, estimated by Cronbach's alpha, was 0.76. According to the analysis of the responses to the question "During the past month, how often have you had trouble sleeping because you _____" in the questionnaire, we received information on the degree of influence of different factors on residents' sleep. The original PSQI only contains options 'A-I' and 'others'. After the statistical analyses, we summarised other factors into 'J-N' as shown in Table 5.1.

Table 5.1 Sleep quality measured by PSQI

Pittsburgh Sleep Quality Index	Bedtime	Sleep latency	Rise time	Total sleep time
Mean	12:37am	40.5min	8:15am	6.8h
SD	45min	36.7min	55min	1h
How often have you had trouble sleeping because ____	Not during the past week	Less than once a week	Once or twice a week	3 or more times a week
A. Cannot fall asleep within 30min	37.78%	42.22%	13.33%	6.67%
B. Easy to wake up or wake up early at night	80%	13.33%	4.44%	2.22%
C. Go to the toilet at night	90%	6.67%	2.22%	1.11%
D. Poor breathing	87.78%	10%	2.22%	0%
E. Cough or snoring	62.22%	26.67%	6.67%	4.44%
F. Feeling cold	65.56%	21.11%	13.33%	0%
G. Feeling hot	60%	20%	20%	0%
H. Nightmare	68.89%	24.44%	4.44%	2.22%
I. Pain/discomfort	90%	5.56%	4.44%	0%
J. Stress/anxiety	28.89%	40%	26.67%	4.44%
K. Coffee, alcohol, staying up late	35.56%	13.33%	28.89%	22.22%
L. Outdoor noise	73.33%	17.78%	6.67%	2.22%
M. Indoor noise	48.89%	26.67%	11.11%	13.33%
N. Light	82.21%	8.89%	6.67%	2.22%
Other reasons	90%	6.67%	1.11%	2.22%

Continued

Pittsburgh Sleep Quality Index	Bedtime	Sleep latency	Rise time	Total sleep time
How often have you ____	—	—	—	—
Taken medicine to help you sleep	100%	0%	0%	0%
Had trouble staying awake during social activities	17.78%	28.89%	31.11%	22.22%
Had a problem getting the enthusiasm to get things done	28.89%	35.56%	26.67%	9%
Rate overall sleep	Very good	Fairly good	Fairly bad	Very bad
Percentage of rate overall sleep	8.89%	62.22%	28.89%	0%
Global PSQI	Optimal (1-5)	Borderline (6-7)	Poor ($\geqslant 8$)	—
Percentage of global PSQI	31.11%	40%	28.89%	—

3. Evaluation of sound environment of sleeping

Participants also needed to finish the designed subjective assessment of sleep sound-environment questionnaire for the previous 30 days after interviews, in which participants were required to list perceived sources of sound in the bedrooms. Sound perception and noise sources for disturbance of sleep were evaluated using an 11-point scale (Pirrera et al., 2014), which was scored from 0 to 10 in three stages—before sleep, during sleep, and after waking up. Four sources of noise were evaluated: outdoor noise (e.g. road traffic, construction/industry, wind/rain/storm/lightning, and construction equipment), noise from roommates (e.g. snores, conversation and whispering), other indoor noises (e.g. footsteps in the corridor and conversation in the corridor), and activities (e.g. activities in the next room, roommates' sleep-related activity, roommates' study and entertainment activities). In the evaluation of the scale of perception, 0 means "completely inaudible", and 1-10 indicate "just heard" to "very significant". While assessing the impact, 0 means "no influence", and 1-10 indicate "slight influence" to "severe influence".

5.1.2.4 Objective measurements

1. Sound-level measurements

Nocturnal environmental noise was recorded inside each bedroom simultaneously for seven consecutive days from Monday to Sunday. A total of 45 rooms were measured for ambient noise using sound-level meters separately, which were placed at ear level of participants in bedrooms to record equivalent sound level (LAeq), peak sound levels [L_{10}, 10th percentiles in dB LAeq], background sound levels [L_{90}, 90th percentiles in dB LAeq] at the measuring point, with slow-style and A-weight (Park et al., 2014) since they are all related to sleep. To better describe the acoustic environment of our study population, we maintained the habitual sleep schedules for the participants as the targeted sleep period studied (Eberhardt et al., 1987). The measurement was carried out in winter, November 2018, for a week. First, the mean noise level value per 10-second period was recorded for 24 hours, for seven consecutive days, and then the sound level of individual sleep environment during sleep latency, sleep and 15 minutes after waking up was calculated according to the measurement of sleep time. Additionally, many previous studies indicate that excessively high or low ambient temperature (Ta) may affect sleep because thermoregulation is strongly linked to the mechanism regulating sleep (Gilbert et al., 2004). In order to avoid the influence of indoor thermal environment on the experimental results, the indoor temperature was measured at the same time. Results from measurements showed that ambient temperatures of the 45 rooms ranged from 22℃ to 23℃ due to the central heating system in the residence hall. Prior research has indicated that a comfortable temperature in winter was close to this range (Lan et al., 2014).

2. Sleep quality measurements

The study was designed to objectively measure sleep in relation to noise exposure using actigraphy, which has emerged as a widely accepted tool to tracking sleep and wake behaviour (Ancoli-Israel et al., 2003; Sadeh, 2011). This method provided a more comprehensive evaluation of the potential effect that noise may have on sleep for college students, when considered together with the self-report. An actigraphy device worn on the wrist was given to all participants to record the sleep quality for a minimum of seven consecutive days. Respondents were asked to wear the device on their wrist during all hours of the day and night for the seven days following their survey (Michaud et al., 2016). The sleep monitoring technology adopted by the wristband is based on an electrocardiogram-based technique named cardiopulmonary coupling technology (CPC; Thomas et al., 2005), which can take exact measurements of sleep patterns, including timing and duration of sleep as well as awakenings (Sadeh, 2011). CPC has been proposed recently as an indicator of sleep stability and as an alternative way of characterising sleep (Thomas et al., 2005), and has been used in some sleep-related studies (Chen et al., 2018; Cysarz et al., 2018; Schramm et al., 2013). Although it cannot replace traditional polysomnography due to imperfect sensitivity and specificity to detect wake periods, this tool can provide reasonable estimates for assessing subjects objectively with minimal participant burden for more prolonged periods of time than conventional assessment tools (Martin and Hakim, 2011).

The actigraphy provided key information on sleep patterns, including total sleep time, deep sleep time, light sleep time, REM sleep time, and the number of awakenings bouts which we used for further analysis. In addition, to help interpret the measured data, respondents were asked to complete a basic sleep log about when they went to bed each night. Then, we got the participants' sleep latency each night also using the wrist data, from which we got sound levels of each participant before they fell asleep, during sleep and after waking up. After the 7-day collection period, respondents were asked to return the completed sleep log along with the sleep watch.

5.1.2.5 Data analysis

Based on the collections of sleep data and questionnaires, the software program SvanPC++ (version 3.3.16) was used to analyse the noise data, and the software SPSS (Feeney, 2012) was used to perform the analyses of the data from the survey. Spearman and Pearson correlation tests were used to calculate the relationship between sound level and influencing factors, the relationship between data of sleep quality and influencing factors, and the relationship between disturbance of noise sources and sleep duration. A t-test at $p < 0.01$ and $p < 0.05$ was used to test for significant differences. Further, linear and nonlinear regression analyses were performed to examine the relationship between noise and sleep duration and between the number of occupants and background sound level among college students. Additionally, an analysis of variance (ANOVA) was used to test for significant differences in self-rated sleep quality between related factors. T-tests were used to test for gender differences, and paired t-tests were used to determine differences between weekday and weekend behaviours.

5.1.3 Results

5.1.3.1 Sleep disturbance from sound environment

Overall, college students reported chronically restricted sleep. Mean total sleep time (time spent actually sleeping, as opposed to being awake in bed) was 6.8h ($SD = 1.0$); bedtime was 00:37 ($SD = 45$min); rise time was 08:15 ($SD = 55$min); and sleep latency was 40.5min ($SD = 36.7$min). Table 5.1 shows responses to individual questions on PSQI. For the 90 participants who completed PSQI in its entirety, the average score was 6.27, with a 95% confidence interval (5.64, 6.89). Of the participants who had poor sleep quality, 68.89% scored greater than 5, and 28.89% scored greater than 8, which means that 68.89% of the college students were poor-quality sleepers. Specifically, 71.11% of students reported lacking the enthusiasm to get things done at least once a week, and 62.22% reported an inability to fall asleep within 30min at least once a week. High rates of insufficient sleep among college students were also found in other related studies (Gilbert and Weaver, 2010; Lund et al., 2010).

Table 5.1 shows that "Stress/anxiety" was the most influential factor, which affected more than 60% of participants' sleep. "Indoor noise" was the third biggest factor after "Coffee, alcohol, staying up late", because of which more than 50% of participants had trouble sleeping. Excessive noise was also a common reason for sleep disturbance in other studies about college students' sleep measured by PSQI. For example, excessive noise was found to be one of the most common "other reasons" at a large private university in the American Midwest (Lund et al., 2010).

In addition, the ANOVA test showed that some personal characteristics had an impact on self-rated sleep quality. Fig. 5.4 shows the result of the question "rate overall sleep", which indicated that self-rated sleep quality is not equal in different position groups and number groups, and higher scores mean poorer sleep, with significance level $p < 0.05$. Participants whose beds were lower bunk or near the door scored higher, which means they sleep worse. This is because there is more corridor noise near the door and more noise from roommate activities on the lower bunk. Besides, groups with greater number of occupants per room rated their sleep quality higher, thinking they have worse sleep. It is noticeable that none of the participants had taken medicine to help sleep in the students sampled, which differs from other studies (Lund et al., 2010).

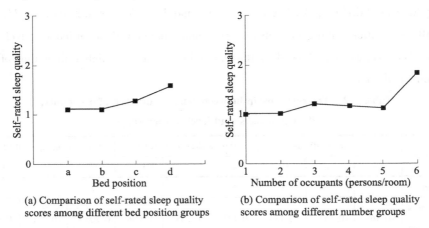

(a) Comparison of self-rated sleep quality scores among different bed position groups

(b) Comparison of self-rated sleep quality scores among different number groups

Fig. 5.4 Comparison of self-rated sleep quality scores among different groups

Note: a. upper bunk beside the window; b. lower bunk beside the window; c. upper bunk beside the door; d. lower bunk beside the door.

5.1.3.2 Effect of number of occupants on nocturnal sound environment

Result of measurement showed that LAeq at night ranged from 21dB(A) to 56dB(A) in the sample, of which LAeq before sleep ranged from 22dB(A) to 56dB(A). LAeq during sleep ranged from 21dB(A) to 42dB(A), and LAeq after waking up ranged from 21dB(A) to 50dB(A). Average night-time sound levels during sleep were the lowest in the three stages, varying within a small range and the highest in the stage before sleep with a large variation range. The peak sound level before sleep reaches a maximum of 65dB(A) and an average of 41dB(A).

As shown in Table 5.2, the number of occupants per room was a significant factor affecting sound level. Pearson correlation confirms a significant relationship between the number of occupants and L_{90}, with significance level $p < 0.05$. There was also a correlation between the number of occupants and L_{10}, the number of occupants and LAeq, with significance level $p < 0.01$. In addition, there was a moderate correlation between floor and sound level, especially during sleep. The higher the floor is, the significantly lower the sound level is. Sound levels also differed significantly by gender in L_{10} [32.3 ±1.7dB(A), 30.6 ±2.2dB(A)] during sleep, L_{10} [36.9 ±4.8dB(A), 33.1 ±4.2dB(A)] and L_{eq} [33.9 ±4.2dB(A), 31.2 ± 4.2dB(A)] after waking up between men and women with significance level $p < 0.05$. The average sound level of the men's bedrooms was higher than that of the women's bedrooms.

Table 5.2 The relationship between gender, number of occupants, floor and the sound level in each stage

Factors	Before sleep (dB SPL)			During sleep (dB SPL)			After waking up (dB SPL)		
	L_{10}	L_{90}	L_{eq}	L_{10}	L_{90}	L_{eq}	L_{10}	L_{90}	L_{eq}
Gender	−	−	−	*	−	*	*	−	*
Number of occupants	−0.169 *	−0.167 **	−0.174 *	0.121	−0.253 **	0.008	0.203 **	−0.441 **	0.100
Floor	−0.051	−0.166 *	−0.007	−0.383 **	−0.201 **	−0.312 **	−0.157 *	−0.033	−0.162 *

Note: ** indicates $p < 0.01$; * indicates $p < 0.05$.

L_{10}—peak sound levels; L_{90}—background sound levels; L_{eq}—equivalent sound levels.

Fig. 5.5 shows the relationship between the number of occupants per room and background night-time sound levels measured with A-weight, along with the corresponding regression curves, the coefficient of determination R^2, with significance level $p < 0.001$. Different regression curves, including linear (R^2 from 0.025 to 0.154), quadratic (R^2 from 0.029 to 0.168) and cubic (R^2 from 0.212 to 0.510), were used to determine the best fit to show the relationship. The value of R^2 with cubic regression in three stages is highest with $p < 0.001$. Thus, this cubic regression is used to explain the relationship between number of occupants per room and background night-time sound levels.

Chapter 5 Sound Environment and Acoustic Perception in Living Spaces

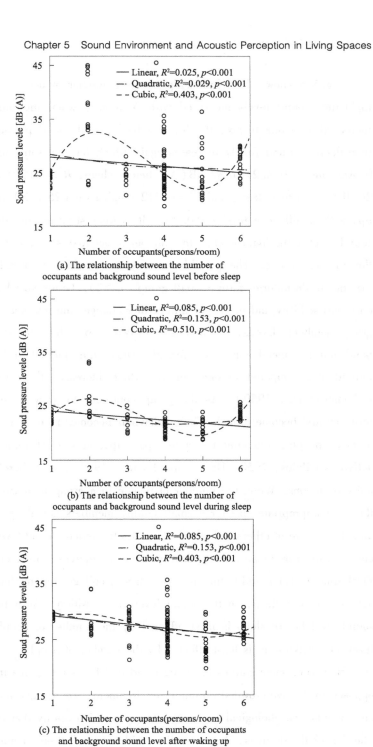

(a) The relationship between the number of occupants and background sound level before sleep

(b) The relationship between the number of occupants and background sound level during sleep

(c) The relationship between the number of occupants and background sound level after waking up

Fig. 5.5 The effect of number of occupants on the sound level in three stages

Fig. 5.5 shows the relationship between number of occupants and background night-time sound levels measured with A-weight: when the number of occupants increased from one to six, the level of background sound pressure first increases, then decreases and finally increases again which has the same tendency with $R^2 = 0.403$ ranging from 21 to 32 dB(A) before sleep; $R^2 = 0.510$ ranging from 20 to 26 dB(A) during sleep; and $R^2 = 0.212$ ranging from 25 to 30 dB(A) after waking up, with significance level $p < 0.001$. It is interesting to note that the background sound level is the highest when the number of occupants is two, and the lowest when the number is five. The possible reasons are as follows. Firstly, students living together in dormitories form a small group ($n \geq 2$). Group size has a certain impact on group stability and intimacy, and further changes the interaction behaviour among group members. Dyads, which consists of two people, have the strongest relationship bond and the membership is informal. Thus, the interaction is focused on both individuals, compared to one person, which increases the level of indoor noise (Simmel et al., 1950). As the group continues to grow in size, members lose intimacy and become formal. Their behaviour becomes more sensitive to the feelings of more people, reducing the active participation rate of related behaviours and activities (Bales, 1950; Darley and Latane, 1968), and therefore, reducing the noise of actions. When the number of people in a group increases to five, which is the most appropriate size for a small group, members get along more harmoniously and think more of others, which also reduces the indoor sound level. As the numbers continue to grow to six, the indoor noise level increases to a certain level again. Furthermore, compared to the other two stages, college students had the most types of activities before sleep, so the sound level varies widely. As a result, the ambient sound level before sleep is most affected by the number of people, and the sound level after waking up is least affected by the number of people.

However, some studies have suggested that fewer occupants in one bedroom, as opposed to the overcrowded dormitory environment, would have an adverse influence on students' psychological and social communication ability that cannot be ignored. The denser the dorm is, the worse the residents rated their roommates (Bickman et al., 1973). A study on the improvement of hospital rooms indicated that patients treated in single rooms were more satisfied with their care than those treated in multiple-bed wards (Lawson and Phiri, 2000). Findings from Fig. 5.5 may help to arrange the appropriate number of occupants, which indicate that when the number is 3, the room with one person is the quietest; and when the number is 4, the room with five persons is the quietest.

5.1.3.3 Effect of sound level on sleep

Table 5.3 provides a summary of the variables retained in the analysis for PSQI and sleep time and relationship between sleep quality and influencing factors. Sound unrelated factors include body mass index (BMI), position of bed and day of the week; and factors related to sound include gender, grade, number of occupants, floor, and noise levels (L_{10}, L_{90}, L_{eq}) at three stages. The results indicate that some factors, such as number of occupants, position of bed, day of the week, noise levels before sleep and noise levels during sleep were significantly correlated with sleep time, with significance level $p < 0.01$, while there was no significant influence on the sleep time by grade, BMI, and noise levels after waking up, where significance level showed $p > 0.10$.

It is necessary to indicate that number of occupants was found to be the most significant factor affecting sleep time; and the more number of occupants are, the less total sleep time ($r = -0.511$, $p < 0.01$), deep sleep time ($r = -0.450$, $p < 0.01$), REM time ($r = -0.503$, $p < 0.01$) and the more awakenings bouts ($r = -0.222$, $p < 0.01$) are, because of its significant effect on sound levels and other behavioural influences. Besides, sleep time differed significantly by day of the week (weekdays or weekends), with significance level $p < 0.01$. Similar results that mean bedtime (01:44am, $SD = 79$min) were delayed and mean rise time (10:08am, $SD = 88$min) were extended on weekends was found in a survey on sleep for college students in an American Midwest college (Van Dongen et al., 2003). College students have longer total sleep time on weekends, mainly due to the course schedules in college, as they do not need to wake up to attend class and instead wake up naturally on weekends. Bed position was also significantly positively correlated with sleep duration, with significance level $p < 0.01$. Living in the lower bunk and near the door affected sleep duration, as it was affected by more roommate activity and corridor noise, which was consistent with results of self-rated sleep quality shown in Fig. 5. 4 (b).

Table 5.3 Relationship between sleep quality and influencing factors

Factors		Total sleep time (min)	Deep sleep time (min)	Light sleep time (min)	REM time (min)	Number of awakenings bouts	PSQI
Sound unrelated factors							
BMI[b]		0.028	−0.262	0.127	0.008	0.092	0.054
Position of bed[b]		−0.267**	−0.242**	−0.095	−0.234**	0.121	0.068
Day of the week[a]		29.668**	9.33*	15.321*	4.556	0.137	−
Sound related factors							
Gender[a]		1.67	12.23*	7.79	1.71	0.36	0.02
Grade[b]		−0.015	−0.038	−0.026	0.035	0.07	0.086
Number of occupants[b]		−0.511**	−0.450**	0.080	−0.503**	0.222**	0.002
Floor[b]		0.105	0.154*	0.007	0.105	−0.182*	0.063
Before sleep	L_{10}^b	0.278**	−0.062	0.244**	−0.299*	0.034	−0.093
	L_{90}^b	0.02	−0.006	0.260**	−0.136	−0.066	−0.05
	L_{eq}^b	0.218**	−0.059	0.266**	−0.221*	0.028	−0.111
During sleep	L_{10}^b	−0.133	−0.034	0.193**	−0.038	−0.003	−0.071
	L_{90}^b	−0.003	−0.382**	0.196**	−0.328**	0.131	0.354*
	L_{eq}^b	0.025	−0.186**	0.229*	−0.121*	−0.04	−0.044
After waking up	L_{10}^b	−0.158	−0.182	0.077	0.155	0.052	0.049
	L_{90}^b	0.122	0.082	0.16	−0.096	0.034	0.127
	L_{eq}^b	−0.091	−0.149	0.099	−0.16	0.03	0.032

Note: ** indicates $p < 0.01$; * indicates $p < 0.05$.

L_{10}, Peak sound levels; L_{90}, Background sound levels; L_{eq}, Equivalent sound levels.

[a] Mean difference.

[b] Correlation coefficients.

As for the effects of sound level, previous studies have suggested that environmental noise causes increased arousal, decreased sleep duration, fragmented sleep, and decreased deep sleep and REM sleep (Clark and Stansfeld, 2007; Muzet, 2007). The relationship in Table 5.3 is similar to results from existing research, especially for deep sleep time and REM sleep time. L_{90} and L_{eq} during sleep were negatively correlated with deep sleep time ($r = -0.382$ and -0.186, respectively, $p < 0.01$) and L_{90} during sleep was also negatively correlated with deep sleep time ($r = -0.328$ and -0.186, respectively, $p < 0.01$). Moreover, light sleep time was found to be significantly positively correlated with sound pressure level before and during sleep, with significance level $p < 0.01$. It is noticeable, however, that L_{10} and L_{eq} before bedtime are positively correlated with total sleep duration ($r = 0.278$ and 0.218, respectively, $p < 0.01$), which is contrary to results from existing research. This occurs for multiple reasons. Sleep deprivation is one of the most common causes of daytime sleepiness among college-aged students. Disturbed by noise before sleep, they go to bed late. If they get up at the usual time, they will not get enough sleep, resulting in sleepiness. Then, some of them will abandon other plans for the morning and continue to sleep to get adequate sleep, which may cause them to sleep longer without any pressure.

Next, regression analyses were used to establish the regression curves of sleep time and noise levels. REM sleep has been identified as a time of learning and memory processing and has been considered in many related studies (Eberhardt and Akselsson, 1987; Kuroiwa et al., 2002). Deep sleep time is an important indicator of sleep quality as it usually occurs early in the sleep cycle and is less affected by total sleep time (Brunner et al., 1990; Ebb and Agnew, 1971). Therefore, we focused on REM sleep and deep sleep time. L_{90} during sleep were found to be significant factors ($r = -0.328$ and -0.382, respectively, $p < 0.01$). Regression curves of REM sleep time and L_{90}, and deep sleep and L_{90} were established, as shown in Fig. 5.6. There have been few studies on the effects of noise below 30dB on sleep before, which was therefore considered an appropriate range (Basner and McGuire, 2018).

Fig. 5.6 (a) shows the relationship between deep sleep time and L_{90} during sleep: R^2 was 0.352, estimated coefficient parameter was -1.7, with significance level $p < 0.001$. The results indicate that deep sleep time takes a linear trend of continuous decrease as the level of sound level increases. When background sound level increased from 20 to 30dB(A) in bedrooms in the residence hall, deep sleep time of college students decreased 1.7min per dB(A). Fig. 5.6 (b) shows the relationship between REM sleep time and L_{90}. Different regression curves, including

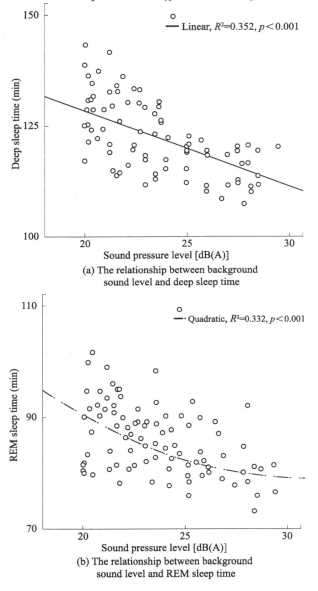

Fig. 5.6 **The relationship between background sound level and sleep time during sleep**

linear and quadratic, were used to determine the best fit to show the relationship, as measured by the value of the coefficient of determination R^2 and statistically significant differences. The value of R^2 with quadratic regression was 0.332, with significance level $p < 0.001$ and linear regression was 0.223, with significance level $p < 0.001$, which indicates that the quadratic regression is the best fit to illustrate the relationship that REM sleep time takes a slow downward trend as the level of sound level increases. Through the estimated coefficient parameter linear curve, REM sleep time may decrease 1.4min per dB(A) as sound level increases.

The relationship between sleep disturbances assessed by PSQI and sound level were also analysed. However, the correlation was not significant, unlike results in extant literature (Basner and McGuire, 2018). Sleep environment in the residence hall is complicated, with many influencing factors; hence, the influence of sound level change is not very significant on PSQI scores.

On the basis of the above results, to improve the sound environment of sleep, the following measures can be taken. Firstly, the dorm bed mode can be adjusted to more loft beds with desks to reduce the influence of roommates' activities. Secondly, the less the number of occupants in each dorm is, the more likely they were to get sufficient sleep. Additionally, sound levels need to be reduced during sleep and before sleep to create a more suitable sleeping environment.

5.1.3.4 Effect of sound sources on sleep

In the assessment, noise sources in the sound environment of sleep in the dormitory were classified as indoor and outdoor noise sources, and indoor noise sources comprised dormitory and other areas. In addition, in order to avoid the impact of some activities, the evaluation of dormitory activities was added to the questionnaire. Perception of and disturbance from noise sources were rated for the month in three stages: before sleep, during sleep and after waking up.

As indicated in Table 5.4, the most common perceived source of noise that caused sleep disturbance was noise caused by conversations of other roommates (77.42%), followed by noise caused by the collision of objects/furniture/door (70.97%), and footsteps in the bedroom (67.74%). Noise caused by roommates' sleep-related activities was the most common source of activities (67.74%).

Table 5.4 Perceived sources of noise by college students in the residence hall

Perceived sources of noise	All ($n=90$)	Number (%) of participants undisturbed sleep ($n=28$)	Number (%) of participants disturbed sleep ($n=62$)
Road traffic	36.67%	35.71%	37.1%
Construction/industry	33.33%	25%	37.1%
Wind/rain/storm/lightning	24.44%	14.29%	29.03%
Construction equipment	26.67%	21.43%	29.03%
Roommate snore	48.89%	57.14%	45.16%
Roommate conversation	71.11%	57.14%	77.42%
Roommate whispery	57.78%	57.14%	58.06%
Footsteps in the bedroom	64.44%	57.14%	67.74%
Keyboard/mouse click	52.22%	53.57%	51.61%
Collision of objects/furniture/door	66.67%	57.14%	70.97%
Footsteps in the corridor	55.56%	50%	58.06%
Conversation in the corridor	53.33%	46.43%	56.45%
Activities in the next room	37.78%	42.86%	35.48%
Roommate sleep-related activity	64.44%	50%	67.74%
Roommate study/entertainment activities	54.44%	57.14%	53.23%

Fig. 5.7 shows the perception of noise and disturbance of noise from various sound sources in three stages. In order to analyse the significance of disturbance from different noise sources, Pearson's correlation analysis was used to calculate the relationship among disturbance of noise sources and sleep time simultaneously.

Before sleep, noise from roommates' conversation, and study and entertainment activities have a significant effect on sleep duration ($r = 0.315$ and 0.322 respectively, $p < 0.01$). As Fig. 5.7 (a) shows, both perception and disturbance evaluation scores of outdoor noise are far less than indoor noise, especially the scores of wind/rain/storm/lightning and construction equipment, which range between 0.5 and 1, which means that participants can hardly hear these two kinds of noises and have not been influenced by them. Besides, most of indoor noises were perceived at roughly the same range, with the scores from 2 to 3. In addition, two types of noise are noteworthy. The effect of keyboard sound on the participants was much greater than the perception of the sound, with a difference in scores of nearly 1. The effect of corridor footsteps on participants was much less than perceived. The results showed that participants were more sensitive to keyboard sounds and less sensitive to corridor footsteps in the acoustic environment before sleep.

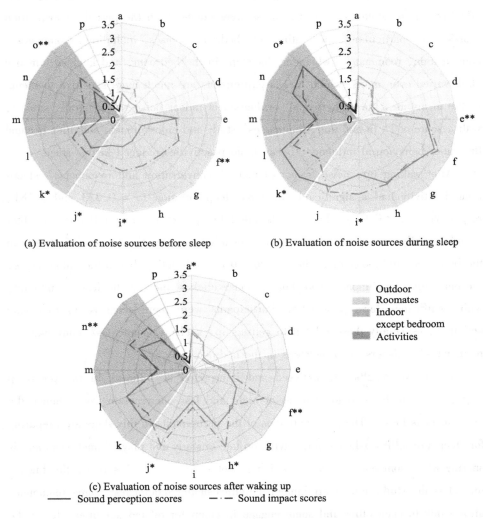

(a) Evaluation of noise sources before sleep

(b) Evaluation of noise sources during sleep

(c) Evaluation of noise sources after waking up
—— Sound perception scores　　—·— Sound impact scores

Fig. 5.7　Evaluation of noise sources in different sleep stages

Note: ** indicates $p < 0.01$; * indicates $p < 0.05$.

a—road traffic; b—construction/industry; c—Wind/rain/storm/lightning; d—construction equipment; e—roommates snore; f—roommates conversation; g—roommates whispery; h—footsteps in the bedroom; i—keyboard/mouse click; j—collision of objects/furniture/door; k—footsteps in the corridor; l—conversation in the corridor; m—activities in the next room; n—roommate sleep-related activity; o—roommate study, entertainment activities; p—others.

During sleep, noise from roommates' snores have a significant effect on sleep duration ($r = 0.35$, $p < 0.01$). Participants were more sensitive to noise, and the effect scores for almost all types of noise were greater than the perceived evaluation scores. In particular, the rating of bedroom noises including roommates' conversation, roommates' whispers, footsteps in the bedroom, and keyboard/mouse click varied even more, which means participants are much more sensitive to them. In the meantime, the participants' evaluation of indoor noise was generally similar, with a score of 2. In addition, roommates' study and entertainment activities became the major behavioural influence, because sleep activities reduced in this stage.

After waking up, noise from roommates' conversation and roommates' sleep-related activity had a significant effect on sleep duration ($r = 0.333$ and 0.281, respectively, $p < 0.01$). Participants rated the perception and influence of other sounds equally, except for roommate conversation, roommate whispers, footsteps in the bedroom, and footsteps in the corridor. It is worth noting that in the indoor noise, the perception and impact of keyboard/mouse click sound are the least significant, with significance level $p > 0.05$. Participants were more sensitive to roommate activities, especially those related to getting up, such as roommates' conversation, roommates' whispers, and footsteps in the bedroom.

The above results are analysed and interpreted as follows. In the pre-sleep stage, there are fewer road vehicles at night and less construction work, hence the noise level is lower. The main activities of the participants in this stage are preparing for sleep (or falling asleep) and staying up late to study and play. Sometimes there is snoring when someone is asleep, which is not common. At this time, the biggest impact is the study and entertainment activities of the roommates. Some roommates always talk to each other and some engage in computer-related activities. It can be seen that the noise generated by these activities has the biggest impact as shown in Fig. 5.7 (a). In addition, due to the poor sound insulation of the door, the corridor noise perceived by the participants is obvious, but the influence was relatively small, because these footsteps and conversations are not persistent. In the sleep stage, with roommates falling asleep, snoring became the main noise, and often lasted for a long time. In the stage after waking up, while most of the students choose to wake up earlier to study in the classroom, there were still some students learning in the bedroom. Other people were either sleeping or performing activities related to getting up and getting enough sleep. The participants who were sensitive to several kinds of sounds were easily woken up. In general, noise perception before sleep is the most significant, followed by noise heard after waking up, and the sounds perceived during sleep are slightly lower, which is the same as the effect analysis. Therefore, arranging uniform schedules and establishing clear regulations on student sleeping behaviours are needed to avoid the interference of others' sleeping. Roommates should reduce some noises including conversation, keyboard/mouse click, and collision of objects/furniture/door. Furthermore, study rooms can be set aside for some students' evening study and entertainment activities in the residence hall, so as not to affect other students' sleep.

5.1.4 Conclusion

Based on the objective measurements and subjective questionnaire survey of sleep quality and sound environment, this study examined the different influences of factors in the sound environment on sleep quality. Based on our findings, several conclusions can be drawn.

First, sound environment was the third biggest factor but the most influential environmental factor among 15 disruptors in PSQI affecting sleep in college students, of which about 50% of participants had trouble sleeping. Second, the number of occupants per room was a significant factor affecting background sound level of sleep, which was the highest when the number of occupants was 2, and the lowest when the number was 5. Other influential factors included gender and floor. Third, sound level is one of the significant influential factors on sleep quality. Total sleep time of college students living in Chinese residence halls was negatively correlated with the peak sound levels, with significance level $p < 0.01$, while deep sleep time and REM sleep time decreased 1.7min and 1.4min per dB(A) as background sound level increased, with significance level $p < 0.001$. In addition, each type of noise had different effects at each stage. The sound environment before sleep is the most complicated and influential, while students were the most sensitive to the noise during sleep. The most common perceived sources of noise disturbing sleep were their roommate conversation (77.42%), followed by collision of objects/furniture/door (70.97%), and footsteps in the bedroom (67.74%).

The present study can help to enhance the sleep quality of Chinese college students through improvements in the sound environment. Firstly, the number of occupants must be adjusted to create a quiet sleeping environment, specifically when the number is 3 or less than 3, the room with 1 person is better; and when the number is 4 or more than 4, rooms with five persons is better. Next, the dorm bed mode can be adjusted to have more loft beds with desks to reduce the influence of roommates' activities. Last, arranging uniform schedules and setting up enough study rooms for some students' evening activities in residence hall are needed to avoid the interference to others' sleep.

The limitations and questions to be further discussed of this study are as follows. First, sleep quality data of whole night were collected by actigraphy. But more detailed sleep data in the specific stages of sleep could not be provided and subtle physiological changes could not be detected. In future studies, polysomnography and other physiological measurements will be used for thorough research. Second, because of the limited functions of noise dosimeters, only the indoor sound level was used in this study as has been done in most studies so far. It has been proposed that noise exposure characteristics play also an import role and that number of events or the peak level of noise events should be analysed which were not available in this study (Basner et al., 2011). In the following study, more detailed noise characteristics such as duration and clarity will be considered and noise events will be recorded, which might provide some insight on the mechanism of noise influenced sleep for college students.

References

[1] AMLANDER C J, FULLER P M. Basics of Sleep Guide. 2nd ed. Westchester, IL: Sleep Research Society, 2005.

[2] ANCOLI-ISRAEL S, COLE R, ALESSI C, et al. The role of actigraphy in the study of sleep

and circadian rhythms [J]. Sleep, 2003, 26 (3): 342-392.

[3] ASTIN A W, ASTIN H S. Sleep deprivation and the development of leadership and need for cognition during the college years [J]. Journal of Adolescence, 2019, 73: 95-99.

[4] AZADBONI Z D, TALARPOSHTI R J, GHALJAHI M, et al. Effect of Occupational Noise Exposure on Sleep among Workers of Textile Industry [J]. Jounrnal of clinical and diagnostic research, 2018, 12 (3): 18-21.

[5] BAKKER R H, PEDERSEN E, VAN DEN BERG G P, et al. Impact of wind turbine sound on annoyance, self-reported sleep disturbance and psychological distress [J]. Science of the Total Environment, 2012, 425 (15): 42-51.

[6] BALES R F. A Set of Categories for the Analysis of Small Group Interaction [J]. American Sociological Review, 1950, 15 (2): 257-263.

[7] BASNER M, MCGUIRE S. WHO Environmental Noise Guidelines for the European Region: A Systematic Review on Environmental Noise and Effects on Sleep [J]. International Journal of Environmental Research and Public Health, 2018, 15 (3): 519.

[8] BASNER M, MÜLLER U, GRIEFAHN, B. Practical guidance for risk assessment of traffic noise effects on sleep [J]. Applied Acoustics, 2010, 71 (6): 518-522.

[9] BASNER M, MÜLLER U, ELMENHORST E M. Single and combined effects of air, road, and rail traffic noise on sleep and recuperation [J]. Sleep, 2011, 34 (1): 11-23.

[10] BICKMAN L, TEGER A, GABRIELER T. Dormitory density and helping behavior [J]. Environment and behavior, 1973, 5 (4): 465-490.

[11] BRUNNER D P, DIJK D J, TOBLER I. Effect of partial sleep deprivation on sleep stages and EEG power spectra: Evidence for non-REM and REM sleep homeostasis [J]. Electroencephalography and Clinical Neurophysiology, 1990, 75 (6): 492-499.

[12] BUYSSE D J, REYNOLDS C F, MONK T H. The Pittsburgh Sleep Quality Index: a new instrument for psychiatric practice and research [J]. Psychiatry Research, 1989, 28 (2): 193-213.

[13] CARPENTER J S, ANDRYKOWSKI M A. Psychometric evaluation of the Pittsburgh Sleep Quality Index [J]. Journal of Psychosomatic Research, 1998, 45 (1): 5-13.

[14] CHEN L D, LIU C Y, YE Z N, et al. Assessment of sleep quality using cardiopulmonary coupling analysis in patients with Parkinson's disease [J]. Brain and behavior, 2018, 8 (5).

[15] CLARK C, STANSFELD S. The Effect of Transportation Noise on Health and Cognitive Development: A Review of Recent Evidence [J]. International Journal of Comparative Psychology, 2007.

[16] CLEGG-KRAYNOK M M, MCBEAN A L, MONTGOMERY-DOWNS, H E. Sleep quality and characteristics of college students who use prescription psychostimulants nonmedically [J]. Sleep Medicine, 2011, 12 (6): 598-602.

[17] CYSARZ D, LINHARD M, SEIFERT G, et al. Sleep instabilities assessed by cardiopulmonary couplinganalysis increase during childhood and adolescence [J]. Frontiers in physiology, 2018 (9).

[18] DARLEY J M, LATANE B. Bystander intervention in emergencies: Diffusion of responsibility [J]. Journal of Personality and Social Psychology, 1968, 8 (4): 377-383.

[19] EBERHARDT J L, STRALE L O, BERLIN M H. The influence of continuous and intermittent traffic noise on sleep [J]. Journal of Sound & Vibration, 1987, 116 (3): 445-464.

[20] EBB W B, AGNEW H W. Stage 4 sleep: Influence of time course variables [J]. Science, 2016, 174: 1354-1356.

[21] EBERHARDT J L, AKSELSSON K R. The disturbance by road traffic noise of the sleep of young male adults as recorded in the home [J]. Journal of Sound & Vibration, 1987, 114 (3): 417-434.

[22] EVANDT J, OFTEDAL B, KROG N H, et al. A population-based study on nighttime road traffic noise and insomnia [J]. Sleep, 2017, 40 (2).

[23] FEENEY B C. A Simple Guide to IBM SPSS Statistics for Version 20.0. Cengage Learning: Boston, MA, USA., 2012.

[24] HERSHNER S D, CHERVIN R D. Causes and consequences of sleepiness among college students [J]. Nature & Science of Sleep, 2014 (6): 73-84.

[25] ISING H, ISING M. Chronic Cortisol Increases in the First Half of the Night Caused by Road Traffic Noise [J]. Noise Health, 2002, 4 (16): 13-21.

[26] GAWLIK K, MELNYK B M, TAN A, et al. Heart checks in college-aged students link poor sleep to cardiovascular risk [J]. Journal of American College Health, 2019, 67 (2): 113-122.

[27] GEORGE C F. Sleep apnea, alertness, and motor vehicle crashes [J]. American Journal of Respiratory & Critical Care Medicine, 2007, 176 (10): 954-956.

[28] GILBERT S P, WEAVER C C. Sleep quality and academic performance in university students: a wake-up call for college psychologists, Journal of College Student Psychotherapy, 2010, 24 (4): 295-306.

[29] GILBERT S S, VAN DEN HEUVEL C J, FERGUSON S A, et al. Thermoregulation as a sleep signalling system [J]. Sleep Medicine Reviews, 2004, 8 (2): 81-93.

[30] GREGORY J M, XIE X, MENGEL S A. Sleep (sleep loss effects on everyday performance) model [J]. Aviation Space & Environmental Medicine, 2004, 75 (3 Suppl): A125-133.

[31] Harbin Institute of Technology. Basic statistics of Harbin Institute of Technology [OL]. http://www.hit.edu.cn/340/list.htm.

[32] KANG S, OU D, MAK C. The impact of indoor environmental quality on work productivity in university open-plan research offices [J]. Building and Environment, 2017, 124 (1): 78-89.

[33] KUROIWA M, XIN P, SUZUKI S, et al. Habituation of sleep to road traffic noise observed not by polygraphy but by perception [J]. Journal of Sound & Vibration, 2002, 250 (1): 101-106.

[34] KWAK K M, JU Y S, KWON Y J, et al. The effect of aircraft noise on sleep disturbance among the residents near a civilian airport: a cross-sectional study [J]. Annals of Occupational and Environmental Medicine, 2016, 29, (9): 9-18.

[35] LAN L, LIAN Z W, HUANG H Y, et al. Experimental study on thermal comfort of sleeping people at different air temperatures [J]. Building & Environment, 2014, 73: 24-31.

[36] LAWSON B, PHIRI M. Hospital design. Room for improvement [J]. The Health Service Journal, 2000, 110 (5688): 24-26.

[37] LEE P J, SHIM M H, JEON J Y. Effects of different noise combinations on sleep, as assessed by a general questionnaire [J]. Applied Acoustics, 2010, 71 (9): 870-875.

[38] LERCHER P, BRINK M, RUDISSER J, et al. The effects of railway noise on sleep medication intake: Results from the ALPNAP-study [J]. Noise Health, 2010, 12 (47): 110-119.

[39] LUND H G, REIDER B D, WHITING A B, et al. Sleep patterns and predictors of disturbed sleep in a large population of college students [J]. Journal of Adolescent Health, 2010, 46 (2): 124-132.

[40] MAFFEI L, IANNACE G, MASULLO M, et al. Noise exposure in school gymnasia and swimming pools [J]. Noise Control Engineering Journal, 2009, 57 (6): 603-612.

[41] MARTIN J L, HAKIM A D. Wrist actigraphy [J]. Chest, 2011, 139 (6): 1514-1527.

[42] MICHAUD D S, FEDER K, KEITH S E, et al. Effects of wind turbine noise on self-reported and objectivemeasures of sleep [J]. Sleep, 2016, 39 (1): 97-109.

[43] Ministry of Education of the People's Republic of China, Statistical Bulletin of National Education Development in 2018. Available online: https://www.moe.gov.cn/jyb_sjzl/sjzl_fztjgb/201907/t20190724_392041.html. (accessed on 24 July 2019).

[44] MUZET A. Environmental noise, sleep and health [J]. Sleep medicine reviews, 2007,

11 (2): 135-142.

[45] ÖHRSTRÖM E, SKÅNBERG A. Sleep disturbances from road traffic and ventilation noise: laboratory and field experiments [J]. Journal of Sound & Vibration, 2004, 271 (1-2): 279-296.

[46] JOLANTA O. Consequences of sleep deprivation [J]. International Journal of Occupational Medicine and Environmental Health, 2010, 23 (1): 95-114.

[47] PARK M J, YOO JEE H, CHO B W. Noise in hospital rooms and sleep disturbance in hospitalized medical patients [J]. Environmental health and toxicology, 2014, 29.

[48] PILCHER J J, HUFFCUTT A I. Effects of sleep deprivation on performance: a meta-analysis [J]. Sleep, 1996, 19 (4): 318-326.

[49] PIRRERA S, VALCK E D, CLUYDTS R. Field study on the impact of nocturnal road traffic noise on sleep: The importance of in-and outdoor noise assessment, the bedroom location and nighttime noise disturbances [J]. Science of The Total Environment, 2014, 500-501 (1): 84-90.

[50] RICCIARDI P, BURATTI C. Environmental quality of university classrooms: Subjective and objective evaluation of the thermal, acoustic, and lighting comfort conditions [J]. Building and Environment, 2018, 127: 23-36.

[51] SADEH A. The role and validity of actigraphy in sleep medicine: An update [J]. Sleep Medicine Reviews, Sleep Medicine Reviews, 2011, 15 (4): 259-267.

[52] SCHRAMM P J, THOMAS R, FEIG B, et al. Quantitative measurement of sleep quality using cardiopulmonary coupling analysis: a retrospective comparison of individuals with and without primary insomnia [J]. Sleep and breathing, 2013, 12 (2): 713-721.

[53] SIMMEL G, KURT H W. The Sociology of Georg Simmel [M]. Glencoe, Ill.: Free Press, 1950: 118-174.

[54] SMITH M G, CROY I, HAMMER O, et al. Vibration from freight trains fragments sleep: A polysomnography study [R], Science Reports, 2016.

[55] SPIRA A P, BEAUDREAU S A, STONE K L. Reliability and validity of the Pittsburgh Sleep Quality Index and the Epworth Sleepiness Scale in older men [J]. Journals of Gerontology Series A, Biological Sciences and Medical Sciences, 2012, 67 (4): 433-439.

[56] STANSFELD S, HAINES M, BROWN B. Noise and health in the urban environment, Reviews on Environmental Health, 2000, 15 (1-2): 43-82.

[57] STRAUCH C, BRANDT S, EDWARDS-BECKETT J. Implementation of a quiet hour: effect on noise levels and infant sleep states [J]. Neonatal network: NN, 1993, 12 (2):

31-35.

[58] SEXTON-RADEK K, HARTLEY A. College residential sleep environmental [J]. Psychological Reports, 2013, 113 (3): 903-907.

[59] THOMAS R J, MIETUS J E, PENG C K, et al. An ECG-based technique to assess cardiopulmonary coupling during sleep [J]. Sleep, 2005, 28 (9): 1151-1161.

[60] THOMAS K P, SALAS R E, GAMALDO C, et al. Sleep rounds: A multidisciplinary approach to optimize sleep quality and satisfaction in hospitalized patients [J] Journal of Hospital Medicine, 2012, 7 (6): 508-512.

[61] THOMÉE S, EKLÖF M, GUSTAFSSON E, et al. Prevalence of perceived stress, symptoms of depression and sleep disturbances in relation to information and communication technology (ICT) use among young adults -an explorative prospective study [J]. Computers in Human Behavior, 2007, 23 (3): 1300-1321.

[62] VAN DONGEN H P, MAISLIN G, MULLINGTON J M, et al. The cumulative cost of additional wakefulness: dose-response effects on neurobehavioral functions and sleep physiology from chronic sleep restriction and total sleep deprivation [J]. Sleep, 2003, 26 (2): 117-126.

[63] WALSH J K, MUEHLBACH M J, HUMM T M, et al. Effect of caffeine on physiological sleep tendency and ability to sustain wakefulness at night [J]. Psychopharmacology, 1990, 101 (2): 271-273.

[64] WEYDE K V, KROG N H, OFTEDAL B, et al. Nocturnal road traffic noise exposure and children's sleep duration and sleep problems [J]. International Journal of Environmental Research and Public Health, 2017, 14 (5): 491.

[65] ZAHARNA M, GUILLEMINAULT C. Sleep, noise and health: review [J]. Noise Health, 2010, 12 (47): 64-69.

5.2 Effects of traffic noise on the sleep of elder adults

5.2.1 Introduction

In addition to air, water and food, the only other biological necessity our bodies require is sleep.[1] Little doubt exists among health professionals about the fundamental importance of sufficient and restorative sleep in maintaining one's physical and mental health. Sleep loss has been implicated in a variety of negative health outcomes[2] including cardiovascular abnormalities,[3] immunological problems,[4] psychological health concerns,[5] and neurobehavioural impairment that can lead to accidents.[6]

It is well established that noise can disturb sleep, and if this disturbance is severe and frequent enough, it can lead to significant fragmentation and sleep deprivation which seriously affects our physical and mental health.[7] Previous studies have shown that residents exposed to environmental noise such as aircraft, road and train noises at night exceeding a certain decibel level suffer from sleep disorders, more or less.[8-13] Among those noises, road traffic noise is the most common and influential factor affecting nighttime sleep for residents. Recently, the World Health Organization's Night Noise Guidelines for Europe[14] suggested a nighttime annual average outdoor level of 40dB(A) to reduce negative health outcomes from sleep disturbance even among the most vulnerable groups.

Elder adults are often troubled by various diseases including insomnia which has important health risk implications. Sleep in elder adults has been extensively studied. Epidemiological studies also indicate a positive correlation between age and the prevalence of insomnia complaints.[15-17] The majority of elder people have reported sleep disturbances[18-20] in many related studies. Therefore, creating a good sleeping environment is very important for elder adults' health. It is important to analyse the various environmental factors that are more likely to affect the sleep of elder people.

However, fewer studies have examined whether elder adults' sleep is more susceptible to traffic noise. Moreover, most of the above studies select European or American cities as research sites; comparatively, there is not enough attention paid to researching pollution hazards of traffic noise in China, which differs in levels of social development and culture. Residents' assessment of the impact of traffic noise is also different from the results in the European and American studies.

Different from urban fringe living areas in Europe and America, Chinese city houses tend toward a high floor area ratio, high-density, and high-rise planning model. As a result, most residential areas are close to the main roads, which may increase the impact of traffic noise on residential areas.

Therefore, taking a typical old-age community in Harbin city, China as the research site and using a questionnaire consisting of the Pittsburgh Sleep Quality Index (PSQI) and other questions, we aimed to reveal the effects of traffic noise on elder adults' sleep.

5.2.2 Methodology

5.2.2.1 Case site

Study locations were drawn from areas in Hongxing Mingyuan Community, Nangang District, Harbin. Harbin is the capital city of Heilongjiang Province, the northernmost province of China. As the economic, political and cultural centre of Heilongjiang Province, Harbin attracts large numbers of people and traffic every day, which causes traffic congestion in the city. There are 21 main roads and 57 secondary roads in the urban area of Harbin. The urban trunk road network density is relatively low, and the road network planning is not reasonable as shown in Fig. 5.8. At the same time, the number of motor vehicles in Harbin is growing annually. The road and bridge facilities cannot keep up with the fast-growing traffic demands of the city. The resulting traffic congestion has led to increasingly serious traffic noise pollution in excess of the national standards in Harbin.

Nangang District, where the Hongxing Mingyuan Community is located, is the central city of Harbin. The residential building is adjacent to the main road (Wenchang Road and Wenchang Viaduct), which has a large amount of traffic every day and is shown in Fig. 5.8 and Fig. 5.9.

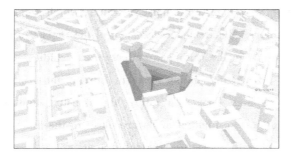

Fig. 5.8　Survey site

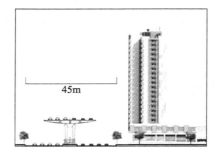

Fig. 5.9　The large road next to the site

5.2.2.2　Participants

There were 83 elder adults aged from 50 to 80 years in the neighbourhood participating in the survey. Of the participants, 52% were women. At the beginning, we randomly selected 110 residents (one person per household) from the neighbourhood. Of them, 93 were willing to participate in the survey, including 10 participants younger than 50 years of age. In the end, we got 83 valid responses.

5.2.2.3　Subject measures

Participants were required to complete a questionnaire including PSQI and other

Chapter 5　Sound Environment and Acoustic Perception in Living Spaces

personal characteristics and residence information.

PSQI is a measure of the subjective experience of sleep that has had detailed psychometric assessment[21] and is one of the most common subjective methodologies used in sleep research. It can provide an assessment of sleep quality over the previous 30 days. PSQI differentiates between "poor-" and "good-" quality sleepers by measuring seven areas: subjective sleep quality, sleep latency, sleep duration, habitual sleep efficiency, sleep disturbances, use of sleep medication, and daytime dysfunction over the past month. [22] The components of PSQI are scored on a scale from 0 (better) to 3 (worse); therefore, the global PSQI is a score ranging from 0-21. A global score greater than 5 is indicative of a poor-quality sleeper, whereas a score of 5 or less is indicative of a good-quality sleeper. [23] The global PSQI score has good internal constancy (Cronbach's $\alpha = 0.71$).

Participants also needed to provide some personal characteristics including gender, age, occupation, family population, and residence information such as floor, bedroom location to the traffic road. Additionally, participants also needed to self-report the impact of traffic noise on sleep using an 11-point scale, which was scored on a scale from 0 (no effect) to 10 (serious impact).

5.2.2.4　Object measures

The traffic noise measurements of the sound pressure level were performed simultaneously. In order to ensure the accuracy and representativeness of the data, the measurement should be carried out under weather conditions without rain or snow, and the wind speed should not exceed 5.5 m/s. The measuring point was selected on the sidewalk 2m away from Wenchang Road. The measurement was carried out in November 2018 for seven consecutive days, from 22:00-8:00am daily. An 801 sound-level meter was set 1m from the wall and main reflectors and 1.2-1.5m from the ground outside the buildings to record LAeq at the measuring point with fast-style and A-weight. We finally got that the equivalent outdoor sound level of the building on the street side is 74.6dB(A).

5.2.3 Results

5.2.3.1 PSQI

According to the calculation method in the appendix, we get the global PSQI score for each participant. For the 83 participants who completed the PSQI, the average score across the entire sample was 5.96 with a 95% confidence interval (5.72, 6.17). Of the participants who have poor sleep quality, 45% scored greater than 5 and 30% scored greater than 8.

The results of the questionnaire also showed that elderly residents go to bed late at night and get up early in the morning, which leads to a low total sleep duration. As indicated in Fig. 5.10, participants go to sleep at 22:12 and get up at 5:36am on average. The average sleep duration is approximately seven hours and twenty-four minutes, which is the optimal sleep duration of 7~8 hours.

According to the statistical analysis of the responses to the question "During the past month, how often have you had trouble sleeping because you _____" in the questionnaire, we got the degree of influence of different factors on residents' sleep. As can be seen from Fig. 5.11, the most influential factor was " c. go to the toilet at night", affecting more than 60% of participants' sleep. About 50% of participants had trouble sleeping because of noise.

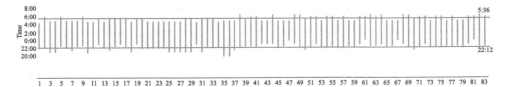

Fig. 5.10 Habitual sleep time and duration

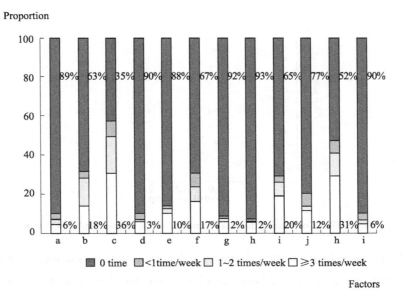

Fig. 5.11 Result of the question "During the past month, how often have you had trouble sleeping because you ___"

a—difficuiltly falling asleep (cannot fall asleep within 30 minutes);
b—easy to wake up or wake up early at night; c—go to the toilet at night;
d—poor breathing; e—cough or snoring; f—feeling cold; g—Feeling hot; h—Nightmare;
i—Pain/discomfort; j—Stress/anxiety; k—noise; l—Other things that affect sleep

5.2.3.2 Effect of traffic noise

From Fig. 5.12, we conclude that noise is the second most important influence on participants' sleep quality, but the specific relationship needs to be further revealed. As can be seen from the scatter plot in Fig. 5.12, PSQI score seems to have a slowly increasing trend with the influence of noise. Upon further data analysis, only a moderate correlation was found to exist between noise effect and self-reported sleep quality (Pearson coefficient = 0.341, $p < 0.01$).

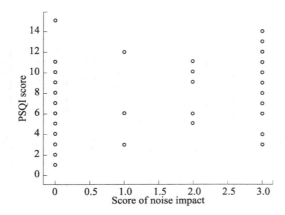

Fig. 5.12 Correlation between score of noise impact and PSQI score

At the same time, another discovery indirectly confirmed the impact of noise on sleep. Participants whose bedroom location is on the side close to the road have higher PSQI scores (6.05, SD = 3.45) than those who sleep on the other side (5.93, SD = 3.82), which means participants whose bedrooms are located away from the road have better sleep quality.

Participants were also required to evaluate the effects of 11 common sources of noise on sleep including road traffic, neighbours, people living with coughing or snoring, family activities, home appliances, elevator equipment, construction/decoration, wind/rain/thunderstorm, community outdoor activities, resident activity beside the road, and others on a 11-point scale scored from 0 (no effect) to 10 (serious impact).

The results showed that the average score of participants' evaluations of the impact of traffic noise is 6.2 points, which is the highest among all the noises. Almost everyone's evaluation score for traffic noise was greater than 0, meaning that it affects almost every participants' sleep.

5.2.4 Conclusions

The survey shows that the nighttime sleep of elderly residents in the neighbourhoods close to roads with a large traffic volume is susceptible to being affected by traffic noise, and the noise impact of self-evaluation is weakly related to sleep quality.

More than 60% of participants' sleep has been affected by having to "go to the toilet at night", making it the most influential factor. The effect of noise is secondary, which is the only environmental factor. Other common factors included "easy to wake up or wake up early at night", "feeling cold", and "pain/discomfort". Therefore, taking measures to reduce environmental noise is the most effective, largely beneficial and simplest method for solving this health problem for most older adults.

The current research still has certain flaws. More samples are needed for further confirmation on the following:

1. We found that gender and family population have a very weak relationship with PSQI. More samples may help us to discover it.

2. We have found that traffic noise has a certain impact on the sleep of elderly residents; however, further quantitative research controlling for noise levels, which will have greater practical implications, may be needed.

3. PSQI is only self-reported sleep quality, which may not be the same as the actual sleep condition. If we can combine some objective measurement methods, the results may be more convincing and interesting.

REFERENCES

[1] GREGORY J M, XIE X, MENGEL S A. SLEEP (sleep loss effects on everyday performance) model [J]. Aviation Space & Environmental Medicine, 2004, 75 (3 Suppl): A125-133.

[2] ZAHARNA M, GUILLEMINAULT C. Sleep, noise and health: review [J]. Noise Health, 2010; 12 (47): 64-69.

[3] SCHWARTZ S W, CORNONI-HUNTLEY J, COLE S R, et al. Are sleep complaints an independent risk factor for myocardial infarction? [J]. Annals of Epidemiology, 1998, 8 (6): 384-392.

[4] JOLANTA O. Consequences of sleep deprivation [J]. International Journal of Occupational Medicine and Environmental Health, 2010, 23 (1) 95-114.

[5] PILCHER J J, HUFFCUTT A I. Effects of sleep deprivation on performance: a meta-analysis [J]. Sleep, 1996, 19 (4): 318-326.

[6] GEORGE C F. Sleep apnea, alertness, and motor vehicle crashes [J]. American Journal of Respiratory & Critical Care Medicine, 2007, 176: 954-956.

[7] BASNER M, MÜLLER U, GRIEFAHN B. Practical guidance for risk assessment of traffic noise effects on sleep [J]. Applied Acoustics, Applied Acoustics, 2010, 71 (6): 518-522.

[8] SCHMIDT F, KOLLE K, KRUEDER K, et al. Nighttime aircraft noise impairs endothelial function and increases blood pressure in patients with or at high risk for coronary artery disease [J]. Clinical Research Cardiology, 2015, 104: 23-30.

[9] KWAK K M, JU Y S, KWON Y J, et al. The effect of aircraft noise on sleep disturbance among the residents near a civilian airport: a cross-sectional study. Annals of Occupational and Environmental Medicine, 2016, 29, (9): 9-18.

[10] NGUYEN T L, YANO T, NISHIMURA T, et al. Social surveys on community response to a change in aircraft noise exposure before and after the operation of the new terminal building in Hanoi Noi Bai International Airport [R]. Paper presented at Internoise, Hamburg, Germany, 2016.

[11] EVANDT J, OFTEDAL B, HJERTAGER K, et al. A population-based study on nighttime road traffic noise and insomnia [J]. Sleep, 2017, 40 (2).

[12] PIRRERA S, DE VALCK E, CLUYDTS R. Field study on the impact of nocturnal road traffic noise on sleep, the importance of in-and outdoor noise assessment, the bedroom location and nighttime noise disturbance [J]. Science of the Total Environment, 2014, 500-501 (1): 84-90.

[13] SMITH M G, CROY I, HAMMER O, et al. Vibration from freight trains fragments sleep: A polysomnography study [R]. Science Reports, 2016.

[14] WHO. Night Noise Guidelines for Europe [R]. Hurtley C (ed). Copenhagen Denmark: WHO Regional Office for Europe, 2009.

[15] KARACAN I, THORNBY J I, ANCH M, et al. Prevalence of sleep disturbances in a primarily urban Horida county [J]. Social Science & Medicine 1976, 10 (5): 239-244.

[16] MCGHIE A, RUSSELL S M. The subjective assessment of normal sleep patterns [J]. The British Journal of Psychiatry, 1962, 108 (456): 642-654.

[17] WELSTEIN L, DEMENT W C, REDINGTON O, et al. Insomnia in the San Francisco Bay area: a telephone survey [M] //C Guilleminault and E Lugaresi, eds. Sleep/wake disorders: natural history, epidemiology, and long-term evolution, New York: Raven Press, 1983.

[18] ANCOLI-ISRAEL S. Insomnia in the elderly: a review for the primary care practitioner [J].

Sleep, 2000, 23 (1): 23-30.

[19] ANCOLI-ISRAEL S. Sleep and aging: prevalence of disturbed sleep and treatment considerations in older adults [J]. Journal of Clinical Psychiatry, 2005, 66 (9): 24-30.

[20] ANCOLI-ISRAEL S, COOKE J R. Prevalence and comorbidity of insomnia and effect on functioning in elderly populations [J]. Journal of the American Geriatrics Society, 2005, 53 (S7): S264-S271.

[21] CARPENTER J S, ANDRYKOWSKI M A. Psychometric evaluation of the Pittsburgh Sleep Quality Index [J]. Journal of Psychosomatic Research, 1998, 45 (1): 5-13.

[22] BUYSSE D J, REYNOLDS III C F, MONK T H, et al. The Pittsburgh sleep quality index: A new instrument for psychiatric practice and research [J]. Psychiatry Research, 1989, 28 (2): 193-213.